Solid-state Circuit Design

USERS' MANUAL

MATTHEW MANDL

RESTON PUBLISHING COMPANY, INC.
A Prentice-Hall Company
Reston, Virginia

Library of Congress Cataloging in Publication Data
Mandl, Matthew.
  Solid-state circuit design users' manual.
  Includes index.
  1. Electronic circuit design—Handbooks, manuals, etc. 2. Semiconductors—Handbooks, manuals, etc. I. Title.
TK7867.M326   621.3815'3'042   76-26174
ISBN 0-87909-784-1

© 1977 by Reston Publishing Company, Inc.
A Prentice-Hall Company
Reston, Virginia 22090

All rights reserved. No part of this book may
be reproduced in any way, or by any means,
without permission in writing from the publisher.

1 3 5 7 9 10 8 6 4 2

PRINTED IN THE UNITED STATES OF AMERICA

Contents

Preface

Chapter 1   Circuit Bias and Other Basics   1
- **1-1**   Summary of Solid-state Physics   1
- **1-2**   Solid-state Junction Structures   4
- **1-3**   Diode Types   5
- **1-4**   Diacs and Triacs   6
- **1-5**   Tunnel Diode   7
- **1-6**   Schottky Diode   9
- **1-7**   Transistor Symbols and Bias   11
- **1-8**   Field-effect Transistors (FET's)   14
- **1-9**   FET Circuit Bias   17
- **1-10**   Basic Common-Source JFET Circuit   18
- **1-11**   Basic MOSFET Circuits   19

Chapter 2   Transistor Characteristics and Parameter Factors   21
- **2-1**   Transistor Characteristics   21
- **2-2**   Other Common-element Basic Circuits   23
- **2-3**   FET Characteristic Curves   25
- **2-4**   Source, Gate, and Dual-gate   26
- **2-5**   Transfer Characteristics   27
- **2-6**   Network Parameters   28
- **2-7**   Hybrid (h) Parameters   33

2-8 Y Parameters   37
2-9   MOSFET Gain Factors   38
2-10 Input-output R Equations   40
2-11 Terminal Identification   40
2-12 Heat Sink Factors   42
2-13 Transistor Testing   44

Chapter 3   Amplifier Circuits   46
3-1   Loss Factors   46
3-2   Load-line Factors   48
3-3   Power Amplifier Load-line Factors   51
3-4   Transformer Factors   56
3-5   Coupling Ratios   57
3-6 Z Matching   60
3-7   Push-pull Matching   62
3-8   Decoupling Circuit   64
3-9   Inverse Feedback   66
3-10 Direct Coupling   67
3-11 Design Procedure Variables and Considerations   68
3-12 Definition of Letter Symbols   69
3-13 Design-related Equations and Applications (Part 1—Junction Transistors)   70
3-14 Design-related Equations and Applications (Part 2—JEFTs)   76
3-15 Design-related Equations and Applications (Part 3—MOSFETs)   81

Chapter 4   R-F Amplification   85
4-1   Circuit Considerations   85
4-2   Resonant Factors Summary   86
4-3   Bandwidth Relationships   89
4-4   Selectivity (Q)   93
4-5 Q Factors (Series and Parallel)   97
4-6   Parallel Relationships   99
4-7   RF Circuits   100
4-8   Neutralization   101
4-9   Common-base RF Amplifier   102
4-10 FET RF Amplifiers   104
4-11 Modulated Class C   106
4-12 AM and FM   108

Contents / v

- **4-13** FM Balanced Modulator 110
- **4-14** Preemphasis and Deemphasis 112
- **4-15** Resonant Filters 113
- **4-16** TV Traps 116
- **4-17** Coil Coupling Factors 119
- **4-18** Mutual Inductance 120
- **4-19** Split stator Tuning Design 123
- **4-20** Pix IF (MOSFET) 129
- **4-21** Varactor Tuning 131
- **4-22** IF Amplifier Filters 132
- **4-23** Hi-fi Terms and Ratings 133

Chapter 5 Oscillator Circuits 138

- **5-1** Crystal Oscillator 138
- **5-2** Variable-frequency Oscillator 141
- **5-3** Tuner Oscillator 142
- **5-4** 3.58 MHz Oscillator 143
- **5-5** Reactance Circuitry 146
- **5-6** Relaxation Oscillator 148
- **5-7** Blocking Oscillator 149
- **5-8** Multivibrator 151
- **5-9** Vertical Sweep Blocking Oscillator 151
- **5-10** MV Vertical System 154
- **5-11** Sawtooth Modification 154

Chapter 6 Special Circuits 158

- **6-1** Flip-flop Circuit 158
- **6-2** One-shot MV 161
- **6-3** RAM and ROM 162
- **6-4** Integrator Circuitry 163
- **6-5** Differentiator Circuit 165
- **6-6** Combined I and D Modifier Circuits 167
- **6-7** Corn-killer Circuitry 170
- **6-8** AM Detection 172
- **6-9** AVC 172
- **6-10** FM Detection 173
- **6-11** TV H-V System 174
- **6-12** Darlington Transistor 176
- **6-13** Differential Amplifier 177
- **6-14** Operational Amplifier 179
- **6-15** Constant-current Source 182

Chapter 7 Integrated Circuits (ICs) and Modules 183
- 7-1 IC Basics 183
- 7-2 IC Component Formation 184
- 7-3 Basic IC Types 188
- 7-4 IC Forms and Diagrams 188
- 7-5 Digital IC Logic 193
- 7-6 *Or* Circuits 193
- 7-7 *And* Gate Circuitry 196
- 7-8 Logic-circuit Combinations 198
- 7-9 Diode-transistor Logic (DTL) 201
- 7-10 Logic Polarity Factors 202
- 7-11 Multiemitter TTL 203
- 7-12 MOSFET *Nand* 205
- 7-13 DCTL Circuitry 206
- 7-14 Commercial IC Packages 206
- 7-15 C-MOS Units 208
- 7-16 Integrated Injection Logic 211
- 7-17 I^2L *Nor* Circuit 213
- 7-18 Schottky Clamp for I^2L 214
- 7-19 Packaged ICs 215
- 7-20 Switching Design Prerequisites 220

Chapter 8 Power Supplies 222
- 8-1 Half-wave Supply 222
- 8-2 Full-wave Supply 224
- 8-3 Bridge Rectifier 226
- 8-4 Voltage Doubling 227
- 8-5 Voltage Tripling 229
- 8-6 Voltage Regulation 230
- 8-7 Zener Diodes 232
- 8-8 Transistorized Shunt Regulator 234
- 8-9 Bleeders and Variable Output Systems 235
- 8-10 SCR Units 239

Appendices
- A Resistor Color Coding 247
- B Capacitor Color Coding 249
- C Transformer Color Coding 253
- D International System of Units (SI) 255

Index 259

Preface

As the name implies, this *Solid-state Circuit Design Users' Manual* is a reference text for those engaged in design practices or related fields in electronics. Hence, numerous equations are included to aid in the acquisition of necessary data for the assembly of particular circuitry. In addition, specific examples are provided when such are indicated to clarify applications of formulas. Finally, numerous basic circuits are illustrated and analyzed to provide the reader with basic information and data regarding their scope, component selection, and utilization of a particular type transistor. This text is not intended to provide a step-by-step teaching guide in design practices, though basic factors necessary for the newcomer in the field are provided in early chapters. Rather, this book aims at providing the individual interested in or already engaged with design programs, with additional data, equations, parameters, and graphs to expedite his work. Toward this end a number of representative schematics of audio, TV, RF, and digital practices have been included, plus discussions on noteworthy commercial types widely used in design practices and integrated circuit devices.

The first two chapters cover a variety of subjects, including the bias factors and requirements for diodes, junction bipolar transistors, junction field-effect transistors (JFET) as well as the insulated-gate field-effect transistor (IGFET). For the MOSFET units, the discussions include both the depletion as well as the enhancement types and representative schematics utilizing these solid-state devices are scattered throughout the text. Network parameters, design equations, and basic circuit configurations are included in the early chapters.

Chapters 3 through 5 cover design aspects and related equations for audio and RF amplifiers as well as low-and high-frequency oscillators. Basic design procedures, load-line factors, and graph interpretations are found in Chapter 3. In Chapter 4 the aspects of selectivity, bandpass, and related design equations are covered. Also in this chapter, special RF circuit diagrams are discussed (such as varactor tuning and ceramic IF filters.) Oscillators and related circuitry are given in Chapter 5. Special circuits are found in Chapter 6 and include AM and FM detection, Darlington and differential circuitry, as well as operational-amplifier factors and equations.

Integrated circuits are covered in Chapter 7, and discussions include basic IC types, digital IC logic systems, and appropriate symbols. Included in this chapter is coverage of DTL, TTL, DCTL, C-MOS, and I^2L integrated-circuit units and applications. Also in Chapter 7 are basic logic tables and illustrations of IC packages such as *or, nor, and, nand,* and *exclusive or* circuits.

Chapter 8 is devoted to power supply systems, factors, and circuitry. Included are discussions of zener and SCR applications as well as voltage regulation and control. The appendix includes color codes and listings of the symbols used in the International System of Units.

MATTHEW MANDL

CHAPTER 1

Circuit Bias and Other Basics

1-1 SUMMARY OF SOLID-STATE PHYSICS

When the atoms of germanium or silicon are brought into close proximity, there is a mutual sharing of outer electrons between any two atoms, which forms a bond having crystal characteristics. The outer-ring electrons that have such bonding characteristics with those of the outer electron shell of adjacent atoms are termed *valence electrons*. In the silicon atom, the outer orbital ring has six electrons.

To move an electron upward into a new orbit, a specific amount of energy must be applied to break the binding effect and to move the electron over a gap and into a new orbital ring. With silicon and germanium, only sufficient effort is required to bridge the space between a subshell ring, because the third subshell of the valence ring is empty.

Electrons that are situated nearest the nucleus of an atom are at low energy levels, whereas those in the outer orbits are at higher energy levels. In a solid structure such as a crystal, the energy levels of individual atoms reform and create *bands of energy levels*. The upper bands of the germanium crystal are shown in Fig. 1-1(a) and consist of the valence-band energy level, the so-called forbidden band, and the conduction energy level band.

In the valence band the electrons are tightly bound within the orbit and hence are not free to carry currents easily. The forbidden band is the gap between the conduction band and the valence band as shown in Fig. 1-1(a). It is this forbidden band (where electrons cannot remain) that must be crossed by the electrons that are to be moved from the

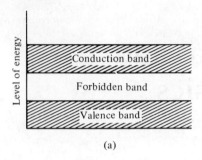

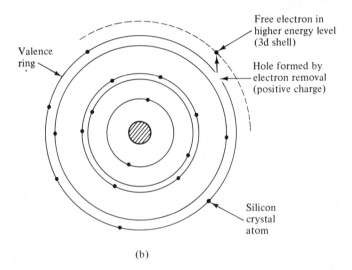

Figure 1-1 Conduction aspects

valence band to the conduction band. The latter band is the one wherein the energy level of the electrons is sufficiently high to permit electron movement and hence current flow.

In materials that have insulating characteristics, a forbidden band is so wide that very few electrons can be given sufficient energy to travel across the gap and reach the conduction band. With the silicon or germanium crystal structure, however, the forbidden band is more narrow, and normal temperature ranges provide sufficient energy for the valence electrons, so they are enabled to reach the conduction band. The quantity of electrons reaching the conduction band depends on the width of the forbidden band, the amount of energy applied, and the temperature of the element. Light rays can also be a source of energy for creating electron movement.

The crystal structures of germanium and silicon are referred to as *semiconductors*. For ordinary conductors such as silver, copper, and aluminum, many free electrons are present at room temperature, and the forbidden region has an extremely narrow (or nonexistent) gap, so that the valence band and the conduction band are virtually one.

When electric pressure in the form of electromotive force is applied to a conductor, the energy pressure causes current flow brought about by electron movement within the structure. Such current flow also occurs in semiconductors when the covalent bond has been broken and an electron of the atom has energy applied to move it into the conduction band. This aspect of an electron leaving the valence band and moving to the conduction band creates a condition peculiar to solid-state devices. This curious condition consists of the formation of so-called *holes* that can, in fact, be considered as current carriers just as electrons are carriers of current. Thus, certain transistors are known as *bipolar* types because of operation with two types of carriers, the electron flow (current) and the hole flow. As described later, some types are unipolar, having only a single charge carrier.

The principle of hole flow is illustrated in Fig. 1-1(b). When an electron leaves the valence band and progresses to the conduction band, it leaves a vacancy in the atom and the latter becomes a positive ion. Since the hole creates a positive area in the orbital subshell from which the electron was removed, the hole can be considered as having a positive charge, just as the electron has a negative charge. The nature of the hole in the atom is such that it can sustain electron flow (current flow) by virtue of the electron and hole movement from one valence band to another, without the electrons moving within the conduction band progression. Assume, for instance, that the free electron of the atom shown in Fig. 1-1(b) has moved on, leaving a positive ion. Now it will be possible for an electron from the valance band of an adjacent atom to break its bond and move into the hole in its neighboring atom. When this occurs, the atom that originally had the hole now becomes a neutral atom again, since it has regained the missing electron. The adjacent atom, however, now has a hole where the electron broke its bond and moved into the hole of the neighboring atom. Thus, even though electrons have moved along (current flow), the energy has been confined to the valence bands because of the hole movement.

For the formation of practical solid-state devices, the pure crystal structure must be modified by the addition of small quantitites of elements. This process consists of adding a controlled amount of so-called *impurities* to the crystal structure. These impurities are actually pure elements but are referred to as impurities only in their relationship to the germanium or silicon crystal structure. The addition of such im-

4 / Circuit Bias and Other Basics

purities (boron, gallium, and indium) form, when added properly to the silicon or germanium, areas referred to as *zones*. There are positive zones (*p* zones) and negative zones (*n* zones), which, when formed into slabs and combined with each other, produce the solid-state devices such as diodes and transistors, as discussed in the next section.

1-2 SOLID-STATE JUNCTION STRUCTURES

When the positive and negative crystal sections called "zones" are joined to form a union, the junction device forms what is known as a *diode* and hence has unidirectional characteristics. Hence, a *p-n* junction has a high resistance in one direction but a low resistance in the opposite direction. Transistors are formed by combining three zones, as explained in the next section.

Potentials applied to solid-state devices are referred to as *bias*, and the two types are *forward bias* and *reverse bias*. The bias relationships for the solid-state diode are illustrated in Fig. 1-2.

The standard symbol for the diode is shown at the center. Note that the side designated as the anode conforms to the *p* zone, while the side indicated as the cathode refers to the *n* zone. When the voltage which is applied conforms in polarity to the two zones, the bias is referred to as forward bias as shown at the left. Thus, for the *p-n* junction, the positive polarity of the battery is applied to the *p* zone and the negative to the *n* zone. If the order is reversed, as shown in the lower left drawing, the battery polarity is also reversed, so each polarity conforms to the respective polarity of the zones. When forward bias is applied to a diode, conduction occurs, and the internal resistance is extremely low.

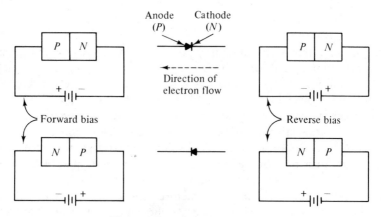

Figure 1-2   Diode bias factors

Reverse bias is shown at the right in Fig. 1-2. Now, a negative battery polarity is applied to the *p* zone and a positive polarity to the *n* zone. Under such a condition no conduction occurs, and internal resistance is extremely high; for practical purposes the diode would be nonconducting up to the limits of the voltage that the diode can withstand without breakdown.

1-3 DIODE TYPES

Diodes have various shapes and come in different sizes, depending on power-handling requirements. Typical kinds are illustrated in Fig. 1-3. For detection purposes where RF-type signals are to be demodulated, the units have the appearance as shown in Fig. 1-3(a), though the diameter may be as small as 1/16 inch. In many of these units the cathode end is marked with a band, as illustrated. For power supply rectification, where currents are high, many diodes are shaped as shown in Fig. 1-3(b), where the bolt arrangement of the cathode side permits mounting directly on a metal chassis or on metal sections that contain heat-dissipating flanges (*termed heat sinks*), as more fully described later. (See also Chapter 8, "Power Supplies.")

In addition to their use in power supplies or detector systems, solid-

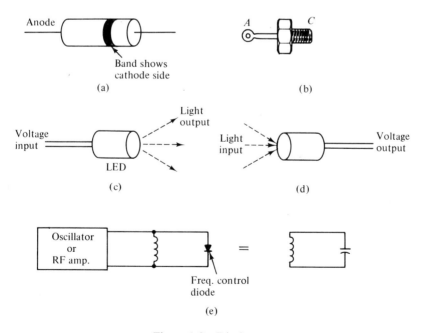

Figure 1-3 Diode types

state diodes have other applications, some of which are shown in Fig. 1-3. One type is the light-emitting diode [Fig. 1-3(c)], where forward biasing of a specially formed *p-n* junction causes a conversion of electric energy to light. The light is produced without the thermal aspects of incandescent lamps, and such diodes are useful for indicator lamps and visual readout devices for test instruments, calculators, and electronic watches.

Reverse function of the light-emitting diode (LED) is shown in Fig. 1-3(d). A photodiode converts light into electric signals by using again a *p-n* junction processed to be sensitive to light photons that strike the junction to create current flow. Such diodes are useful for sensing devices in industrial control, alarm systems, and reading devices in computer input applications.

Figure 1-3(e) shows a diode in which the junction capacitance can be changed by applying a voltage, and thus creating a variable-capacitance diode of the *p-n* type. Such diodes are also referred to as *varactors;* they are widely used in electronic circuitry (such as television receivers) for tuning without mechanical means. Such diodes can also be used for establishing the frequency of signal-generating oscillators. (See Sec. 4-21.)

The symbols for the special diodes are modified from the basic type shown in Fig. 1-2 and appear as shown in Fig. 1-4. At (a) is the variable-capacitor-type diode, while at (b) is the type known as the *zener*. The zener is a voltage-reference diode useful for voltage regulation, as more fully discussed in Chapter 8. At (c) is shown the photodiode (sometimes referred to as a *solar cell*). The LED is shown at (d). At (e) is the silicon-controlled rectifier (SCR) sometimes referred to as a *thyristor*. This diode is useful for switching current flow by use of the gate. (See Chapter 8, Sec. 8-10.)

1-4 DIACS AND TRIACS

The diode shown at (f) of Fig. 1-4 is termed a Diac and contains a dual arrangement of diodes side by side, but having opposite polarities in their parallel arrangement. Such a device is used as a special gating unit because of its characteristic of not conducting until a specific breakdown voltage is applied. When conduction does occur, it will pass current in either direction. Diacs are rated at specific breakover voltages. For instance, if a Diac is classified as 6 volts, it will appear as a high resistance with virtually no current flow for voltages less than 6 volts. As soon as the 6-volt breakover potential occurs, both diodes then have the ability to conduct.

With the addition of a gate lead, as shown at (g), the unit forms

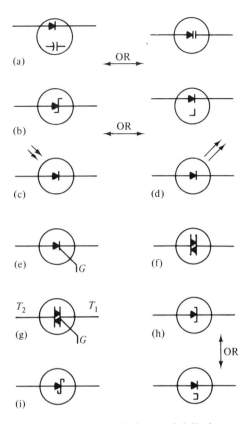

Figure 1-4 Symbols for special diodes

what is known as a Triac, and is a three-terminal unit as shown. The unit is an electronic gate or switch similar in function to the silicon-controlled rectifier. Unlike the latter, however, the Triac has the ability to conduct in both directions and hence can handle ac signals.

As with the SCR, the Triac's gate input lead is for triggering purposes. Since current flow is in either direction, the terminals are not designated as cathode and anode with their implications of polarity. Instead, the terminals can be numbered for reference, since they can be interchanged in the series circuit. It is the gate element that is the important one for triggering purposes. As with the SCR, a specific voltage must be applied to the gate before conduction occurs. (See Chapter 8.)

1-5 TUNNEL DIODE

The diode shown at Fig. 1-4(h) is termed a *tunnel diode*. The tunnel diode has negative-resistance characteristics and hence is useful for rapid signal-switching purposes, as well as for special gating applications. Unlike the standard silicon diodes, the primary application for the tunnel diode is not rectification, but the special high-speed gating and switching mentioned. Switching applications are more rapid than junction transistors by a ratio of over one hundred to one.

The tunnel diode does not have the sensitivity to changes of temperature to the degree found in the average transistor, and consequently operates under stable conditions for wide temperature changes. Also, the tunnel diode resists the adverse effects of nuclear radiation and thus finds wide applications in this field. Radiation effects in ordinary semiconductor diodes and transistors cause changes in internal resistance and also increase noise levels, which may interfere with signals being handled.

The term *tunnel diode* stems from the so-called "tunnel effect" that occurs between the *p-n* junction because of the very narrow barrier formed during the addition of the impurity elements. In some processes, impurity elements are in excess of the amounts normally employed for the junction transistor. Hence, an electric particle reaching the barrier suddenly disappears and reappears almost instantly at the other side of the barrier. The transfer occurs at the speed of light (approximately 186,000 miles per second), as contrasted to the transistor, where charges move much more slowly. Such action appears as though the particle "tunnels" under the barrier rather than going through it. Thus, the tunnel diode has operational capabilities at signal frequencies extending into the gigahertz range.

The tunnel diode can operate as an amplifier because of the negative-resistance characteristics. As shown in Fig. 1-5(a) the diode has a rise in current when forward bias is applied. Such current increase, however, from the zero point (L) will soon reach the peak marked M. An additional voltage increase will now drop current amplitude from the M level to N, as shown. Such a decrease in current represents negative resistance because of the rise of internal resistance with an increase in applied voltage. Any additional increase in potential will cause a gradual rise from N to O, as shown.

When a fixed forward-bias potential is applied so that the operating point is set between N and M, amplification can be obtained. An input sine wave, for instance, will alternately add and subtract from the forward bias value and cause a comparatively high signal-current change to occur within the diode. As shown in Fig. 1-5(b), if the input signal applied to the basic circuit is such that it increases the forward bias, current decreases and lowers the voltage drop across R_2, producing a negative-signal output change.

For an input signal opposing the forward bias (and thus reducing

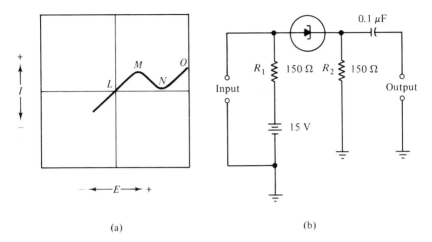

Figure 1-5 Tunnel diode graph and circuit

the amplitude of the fixed potential) there would be a rise in current. Consequently, a positive-polarity signal change occurs at the output of the circuit. Thus, signal amplification is achieved.

1-6 SCHOTTKY DIODE

Figure 1-4(i) shows the symbol for a special diode termed *Schottky diode*. This unit is characterized by rapid switch-on and switch-off times far superior to the junction diodes. In addition, because it has a lower voltage drop than do the junction types, the Schottky has much less power loss and hence is widely used in integrated circuitry for digital applications (see Chapter 7). Basically, it uses an *n*-type solid-state material as the building block (*substrate*) with a metallization-type deposit of material such as molybdenum. Interaction between the substrate and the deposition forms a significant energy barrier, which halts current flow until a sufficient amplitude of bias is reached. For the power Schottky the forward voltage may drop to 0.7 V or less for a 100-A current passage, whereas for a comparable junction diode the forward-voltage drop may reach 1.25 V. Hence, the Schottky diodes are also very efficient in high-speed gating and switching circuitry. Once current flow is established in the Schottky, it is limited only by the unit's semiconductor internal resistance.

For the Schottky no positive-charge hole flow occurs; thus the device is a majority carrier utilizing only electron flow. Hence, recovery time is limited only by its internal capacitance (which stores charges during reverse bias) but is still better than the junction diode, which has

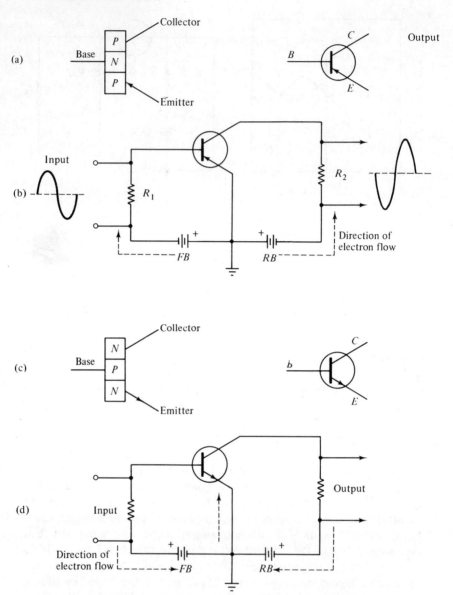

Figure 1-6 Basic transistor symbols and circuits

a recovery time limited by minority-carrier (hole) charge storage. Thus, for equivalent circuit characteristics and potentials, the Schottky diodes will have a better recovery time by a ratio of over 100 to 1 compared to junction diodes.

1-7 TRANSISTOR SYMBOLS AND BIAS

Triode (three-element) transistors are formed by combining three zones, as shown in Fig. 1-6(a), which also shows the symbol for this transistor with corresponding elements identified.

Note that the element connected to the center of the three zones is called the *base*. One of the terminals to a *p* zone is called *collector*, and the terminal going to the other *p* zone is designated as *emitter*. Such a transistor is a *pnp* type, and the arrow that represents the emitter terminal points toward the symbol junction as shown. When the junction is formed as shown in Fig. 1-6(c), where only a single *p*-zone is sandwiched between the two end zones, the transistor is designated as an *npn* type. The symbol for this unit is shown also in (c). Note that now the emitter terminal has the arrow pointing away from the symbol junction to denote an *npn* type.

A basic amplifier circuit is shown in Fig. 1-6(b) to illustrate the application of *forward* and *reverse* bias. This type of circuit is known as a *grounded emitter* (or *common emitter*), because the emitter is placed at direct ground as shown or at signal ground, as illustrated later in Fig. 1-7(a). For Fig. 1-6(b) resistor R_1 in series with the battery makes the base of the transistor negative with respect to the emitter, thus establishing a forward bias, as discussed earlier for the *pn* diode. For most transistors, absence of forward bias or the application of a reverse bias between the base and emitter would inhibit conduction.

Resistor R_2 in series with the battery at the output establishes the proper bias between collector and emitter. Since the bias potential at the collector is minus with respect to the plus emitter, this constitutes a reverse bias, because the collector is a *p* zone and for forward bias would have a positive polarity applied to it. With the reverse bias at the output, the signal is developed across resistor R_2 and represents an amplified version of the signal applied to the input. As shown, when a sine wave signal is applied across R_1, the amplified version appears across R_2 but with a 180-degree phase difference. Adjustment of the bias polarities both at the input and output determines the type of characteristic established for an amplifier. When some idling current flows at all times and the bias is set for the linear portion of the characteristic curve, class A amplification is achieved. For class A, however, the input signal has an amplitude sufficiently low to keep distortion at a minimum.

For class B operation, the forward bias would be reduced or reversed so that the transistor conducts only for one alternation of each sine wave cycle applied. Thus, push-pull operation is required unless an RF amplifier is desired. (See Chapter 3.)

12 / Circuit Bias and Other Basics

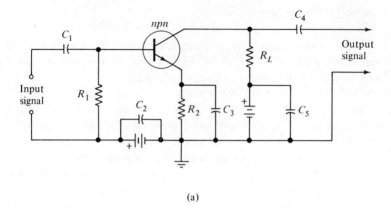

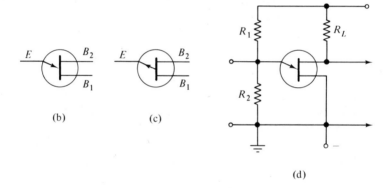

Figure 1-7 Grounded-emitter circuit and UJTs

For class C operation the bias would have to hold the transistor sufficiently beyond the conduction cutoff so that only portions of one alternation cause signal current between emitter and collector, as also detailed more fully in Chapter 3.

For the circuit shown in Fig. 1-6(b), under class A operating conditions, a positive alternation appearing across R_1 would subtract from the bias potential and thus have the effect of reducing the forward bias. Consequently, the reduction in forward bias causes a drop in the conduction existing between the collector and the emitter. In consequence, the voltage drop across R_2 declines, producing an amplified version of the input alternation, except for the 180-degree phase difference. When the input alternation is negative, it aids the forward bias, thus increasing the total forward bias potential. Conduction between collector and emitter increases, a larger voltage drop appears across R_2, and a positive output alternation is obtained as shown.

When an *npn* transistor is used such as shown in Fig. 1-6(d), operation is identical to that described for the circuit in (b), except for a reversal of the battery polarities, as shown. Now, the battery must apply a positive potential to the base in relation to the emitter for the necessary forward bias for the *n-p* input zones. On the collector-emitter side, only a positive battery potential is applied to the collector (which is a negative zone), thus establishing the reverse bias. Again, there is a 180-degree phase difference between input and output signals.

The circuits shown in Fig. 1-6 are basic types illustrating fundamental principles. In practice, however, other components must be added for optimum operating results. Typical is the circuit shown in Fig. 1-7(a). Here, the base input is isolated from the device used to apply a signal by using capacitor C_1. This capacitor may have a value ranging from 0.5 μF to 50 μF. Capacitors C_2 and C_5 are employed to bypass the signals around the power system or batteries as shown so that the internal battery resistance does not add to or disturb the required load resistance (R_L). The signal developed at the output is coupled to the next stage, capacitor C_4 being used as shown. This also has a value within the range specified for C_1.

Resistor R_2 and capacitor C_3 tend to stabilize operation for the transistor, particularly during warm-up. Capacitor C_3 bypasses signal energy across R_2. Current flow through R_2 sets up a voltage drop that is plus at emitter and negative at ground. Consequently, the voltage drop across R_2 will alter the forward bias, since the voltage across R_2 is in opposition to the input bias battery, and the base potential. Similarly, the voltage drop across R_2 is inverse to the output battery; hence the reverse bias is also altered. A reduction in both the forward and reverse bias will reduce the conduction through the transistor, thus compensating for thermal effects. (Other reduction of thermal effects is accomplished by heat sinks, as more fully described later.)

Again, a *pnp* transistor could be used instead of the *npn* in Fig. 1-7. Operational characteristics would be identical, except that the battery potentials would be reversed so the proper forward bias would be established at the input and the reverse bias at the output. (With reverse bias applied to both input and output circuits, there would be no conduction. If forward bias were applied to both input and output sections, however, excessive currents would circulate.)

Unijunction transistors (UJT) were the forerunners of the FET units described next in Sec. 1-8. The UJT device was formed from an *n*-type silicon slab to which another element was bonded (such as indium) to form a *p-n* junction. As shown in Fig. 1-7(b) and (c), the UJT has two base leads and one emitter. The basic circuit showing voltage polarities is given in Fig. 1-7(d). Here resistors R_1 and R_2 form a voltage divider across the power source; by setting the ratio of the two resistors, the proper bias can be applied to the emitter. Resistor R_L is the load resistor across which the output signal develops.

The n-type base symbol is shown in Fig. 1-7(b), and the p-type base symbol is shown in (c). In (d) a p-type could also be used with proper polarity reversals of the supply potentials. When a voltage is applied to the emitter in reverse-bias form (or zero bias) the unit behaves as a conventional resistor. Because the emitter terminal comes in contact with the silicon-slab foundation, a voltage difference is established between emitter and ground. With a forward bias applied to the emitter, transistor characteristics are formed and the resistance of the slab between emitter and Base 1 decreases, so a current increase results. The emitter voltage can now be decreased with a consequent current increase (negative resistance), as was the case for the tunnel diode described in Sec. 1-5. Because a reduction of emitter potential in the UJT no longer causes a decrease in base current, the unit has some of the gating characteristics of the SCR. The UJT can be used to form oscillators or other kinds of circuits. In many applications, however, the more modern solid-state devices described later herein have replaced the UJT.

1-8 FIELD-EFFECT TRANSISTORS (FETs)

The so-called field-effect transistor (FET) differs to a considerable degree from the bipolar junction transistor, because the FET is a *unipolar* device having only a single charge carrier. The charge carrier can be *either* (but not both) electron flow or hole flow, depending on type.

Just as with bipolar transistors, the FET units are available in two polarity types, but instead of *npn* or *pnp* designations the FET are termed n-channel and p-channel units. Current carriers in the n-channel types consist of electrons, and the carriers for the p-channel types are the so-called *holes* formed by electron removal.

In contrast to the bipolar transistor the field-effect transistor has high input and output impedances (comparable to vacuum tubes), and is suitable for low- or high-signal circuitry. The FET also lends itself to a variety of amplifying and switching circuitry (both in the audio and RF ranges), and has excellent thermal stability. In addition, the superior linear characteristics of the FET reduce its susceptibility to cross modulation, where such signals as carriers intermingle to create signal distortion. Thus the FET devices are particularly useful in RF stages of TV and radio receivers for improving performance.

The field-effect transistors are also more capable of processing greater amplitude variations of the input signal before reaching the limits between current cutoff and saturation. Hence, they accept control by AGC better than common transistors where nonlinear conditions often occur. For RF operation the FETs hold internal capacitances close to stable values as opposed to the bipolar transistor. Thus a better RF selectivity as well as stable input and output impedances of the FET are maintained.

The control of electron flow, and hence of current within the FET, is primarily by the influence of electrostatic fields within the MOSFET. Hence, the input signal need not furnish power by being required to supply current to the input of the MOSFET. For the junction bipolar transistor, input power requirements are common. The bipolar transistor is basically a signal *current* amplifier, reproducing at the output an amplified current replica of the signal current change at its input circuitry. For the FET, a signal *voltage* applied to the input *controls* the reproduction of the amplified signal current output.

There are two basic types of field-effect transistors, the *junction type*, which has for its symbol the capital letters JFET, and the metal-oxide semiconductor type, with the symbol *MOSFET*. The MOSFET is also known as the insulated-gate field-effect transistor (IGFET). Both types are fabricated to have either the n-channel or the p-channel operational characteristics.

Symbols for the basic types of FET units are shown in Fig. 1-8. In (a) the two channel types of the JFET are illustrated. The bipolar transistor terminals of base, emitter, and collector are not used here. Instead, the terminals are referred to as *gate, source, and drain*. The gate can be compared to the base of the bipolar transistor, the source terminal to the emitter, and the drain to the collector. For the p-channel the arrow points away from the vertical connecting section as shown, whereas for the n-channel the arrow points toward the vertical section. Special types are shown in Fig. 1-8(b), where an additional gate (G_2) is used. Again, either p or n types are available. The second gate is actually isolated from the first gate. The additional gate forms what is termed a *double* or *dual-gate* FET for circuitry requiring additional signal inputs.

For the MOSFET in (c) and (d) of Fig. 1-8 the gate electric field is shown as a capacitance-coupled device, and not as an arrowhead showing a biased-junction symbol. The capacitance is formed by the deposition of a very thin insulating layer of oxide separating the insulating gate metallic contact. It is from this fabrication process that the term *metallic-oxide field-effect transistor* originates.

For the MOSFET symbols in (c), the n-channel terminal-depletion types are shown. The element marked SUB indicates the substrate ele-

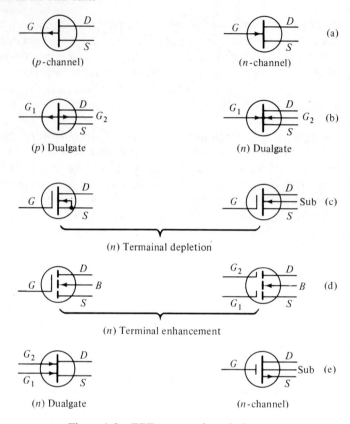

Figure 1-8 FET types and symbols

ment terminal. For *p*-channel types, the arrow would be shown in the reverse direction. The *n*-channel terminal enhancement types illustrated in (d) are identified by having separate segments for the drain, source, and sub, as shown. More details on the significance of these elements are covered in the next section, where bias factors are considered.

As briefly outlined in Sec. 1-9, the foundation slab used in the formation of the FET is termed a *substrate,* and the terminal connecting to this section is sometimes indicated as *sub* or B for bulk.

During the early years of field-effect transistor development there was no general industry standard for symbols, and often different types were used by various manufacturers. In some instances slanting lines were used for the gate terminal or even for the source and drain elements. Some of these early representations will still be found in the literature. Some manufacturers still use alternative symbols and may represent the two gates illustrated in Fig. 1-8(d) as shown in (e), where an alternate representation of the *n*-channel terminal-depletion type is also given.

1-9 FET CIRCUIT BIAS

The junction field-effect transistor (JFET), *n*-channel type, has a lower channel conduction when a *negative* gate voltage is applied, because such a voltage *depletes* the channel of carriers within the FET. For the *p*-channel FET, however, a *positive* gate voltage decreases conduction. The JFET is a normally "on" device, because a conductive channel exists between the source and drain elements, and hence current has a path even without a gate-to-source voltage, V_{gs}. (The channel can actually conduct current in either direction, drain to source or source to drain, because current flow is a function of V_{gs}.) Thus, the JFET, whether an *n*- or a *p*-channel, always has the input element junction operating point established by *reverse bias* in contrast to the *forward* bias applied to the input of the bipolar solid-state units.

In the MOSFET the gate is insulated from the channel by a dielectric, such as silicon dioxide. Thus, compared to the JFET, the gate can have forward bias to enhance the channel as well as reverse bias to deplete it. Thus, two different types of MOSFETs can be designed, the *enhancement* and the *depletion*.

The depletion-type MOSFET is designed to have drain-current flow in the absence of bias at the input. When bias is applied, the drain current is decreased (depleted) to a value necessary for signal-handling (dynamic) operation. As shown in (c) of Fig. 1-8, the symbol line connecting the gate, drain, and base elements is unbroken and can be considered as depicting a closed circuit permitting current to flow until it is adjusted to the operating value by the application of bias.

The enhancement-type MOSFET is designed so that there is little drain current flow in the absence of bias at the input. When bias is applied, drain current is increased (enhanced) to that required for dynamic (signal-handling) operation. As shown in Fig. 1-8(d), the symbol lines for gate, drain, and source elements are separate and thus represent open-circuit conditions, which require the application of bias to bring drain current up to that required for proper operation.

Thus, the MOSFET has three operational modes, depending on the manufacturing process. One is by the application of reverse bias (depletion mode) often referred to as type A. Another method is with 0 bias, the depletion-enhancement mode termed type B. The third method is with forward bias, the enhancement mode, called type C. Generally the JFET units also operate in the A mode.

Basically, in the fabrication of the MOSFET, a substrate (founda-

18 / Circuit Bias and Other Basics

tion slab), such as an *n*-type silicon slice, is used and a *p*-type source and drain regions are diffused into the slab by various masking devices. With the junction FET, the gate is made up of a *p-n* junction diffused into the channel material. The polarities established during manufacture in relation to the materials used determine whether the FET is a *p*-channel or an *n*-channel type. (Additional fabrication summaries are given in Chapter 7 in relation to integrated circuits.)

For the MOSFET the gate insulation is extremely thin, and some units can be damaged by gate insulation puncture caused by static electricity. Thus, such vulnerable types are packaged with precautionary devices (such as shorting wires across leads) to minimize damage by electrostatic fields before installation. Some MOSFETs are of the gate-protected type containing a built-in diode. With the nonprotected types zener diodes are sometimes used in the associated circuitry for voltage-surge protection. (This is illustrated in Sec. 1-11.)

1-10 BASIC COMMON-SOURCE JFET CIRCUIT

The basic common-source JFET circuit is shown in Fig. 1-9. This compares to the grounded-emitter circuit shown in Fig. 1-7. As with that circuit, there is a phase reversal between the input signal and the amplified version appearing at the output.

As with the junction transistor circuit, the input is applied across a resistor (R_1) and hence appears at the gate terminal of the FET. The

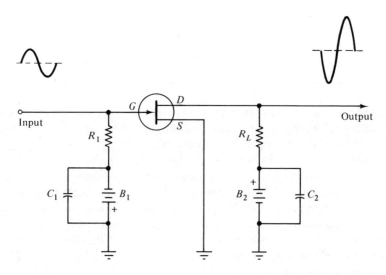

Figure 1-9 Basic circuit of common-source JFET

source terminal is grounded, although, as with the circuit shown in Fig. 1-7, a series resistor and bypass capacitor can be employed for stabilization purposes. The drain element of the FET in Fig. 1-9 is used for the output circuit, and the amplified signals appear across the load resistor R_L, as shown. As with the bipolar transistor circuits, each battery supply (or power supply if used) is adequately bypassed, so that any internal resistance of the batteries will not add to and modify the characteristics supplied by R_L. [Additional circuitry for the JFET (as well as for other transistors) is covered in subsequent chapters.]

1-11 BASIC MOSFET CIRCUITS

Gate-protecting diodes for the dual-gate *n*-channel terminal depletion MOSFET are shown in Fig. 1-10(a). Here, zener diodes are used (see Chapter 8, Sec. 8-7); these are D_2 and D_3. They regulate and limit the voltages applied to the gates and thus protect them from overloads.

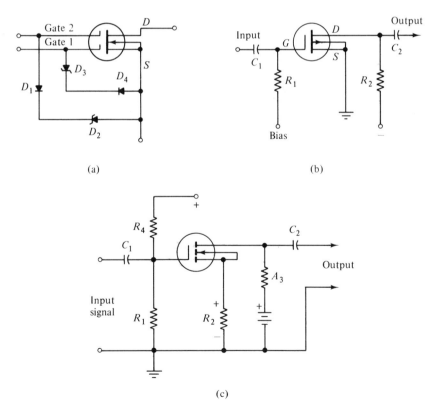

Figure 1-10 Basic MOSFET circuitry

Because the zener diodes function as voltage regulators when reverse-biased, diodes D_1 and D_4 are used to establish a one-way current-flow path (and hence are referred to as *steering* diodes).

A *p*-type terminal-depletion MOSFET is shown in Fig. 1-10(b). Note that the signal input for this common-source amplifier is applied to the gate, as with the JFET of Fig. 1-9, and the output is obtained from the drain by using coupling capacitor C_2. A negative potential is applied to the drain as shown.

For the circuit in Fig. 1-10(c), an *n*-type MOSFET is shown, with resistor R_1 placed at ground. In this position the bottom of R_1 is common to the bottom of the source resistor R_2. The voltage drop caused by the source-drain current develops a polarity across R_2 as shown, thus making the gate input more negative with respect to the source potential. Thus, resistor R_2 alters the bias for the gate-source input, as was the case for the circuit shown in Fig. 1-7. (See also Secs. 2-4 and 3-15.)

A capacitor can be used across R_2 to minimize signal voltage variations. If omitted, some degeneration will result, because signal variations across R_2 are inverse to the amplified signals developed. Thus a form of current feedback occurs which, though it reduces signal gain, aids in reducing distortion and also improves the frequency response of the amplifier. A bypass capacitor has a different reactance for signals of different frequencies. Hence, high-frequency signals are shunted across the resistor, and no degeneration results. Low-frequency signals, however, find a higher capacitive reactance and hence are not completely shunted around the resistor. Consequently, such signals will be attenuated because of the degenerative feedback. Thus, the low-frequency response of the amplifier is diminished. In view of this fact, a flat frequency response is difficult to achieve when the bypass capacitor is used.

CHAPTER 2

Transistor Characteristics and Parameter Factors

2-1 TRANSISTOR CHARACTERISTICS

Characteristic curves for transistors can be graphed by applying certain voltages and taking measurements, or a curve plotter can be used that will automatically establish a set of characteristic curves and display them on a oscilloscope. Characteristic curves are useful for obtaining specific information concerning the voltage ranges, operating characteristics, and other data for design of specific circuits.

Changes in the input voltage to the transistor amplifier can be made, while the changes in the base current that occur in consequence are noted. Thus, voltage may be applied to the collector side and the collector currents may be graphed for various changes of base current. Collector voltages are changed, and, again, collector currents are graphed with respect to base currents. When the various values are set down in graph form, we obtain a set of characteristic curves like those shown in Fig. 2-1. (These curves relate to the common emitter circuit discussed in Chapter 1.)

From the curves shown in Fig. 2-1, the current gain of a grounded-emitter amplifier circuit can be obtained. In such a circuit, current gain is known as *beta,* and the symbol for it is the Greek lowercase letter β. Current gain, in such an instance, refers to the *signal* current gain that results from a change of signal current in the base-emitter side, with respect to the amplified version of the signal current in the collector-emitter side.

As an illustration of the method employed for ascertaining current

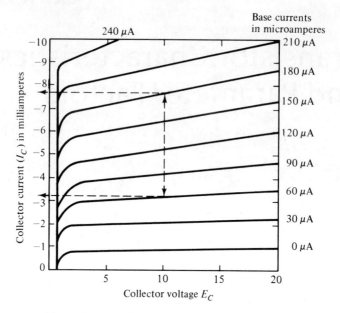

Figure 2-1 Typical grounded-emitter curves

gain for the curves in Fig. 2-1, assume that the collector battery voltage is held at 10 V. (The battery voltage at the base side is not a factor here, since only the signal current *changes* in the base side need be considered.) If a signal voltage is now applied to the base-emitter circuit to cause a change of base current from 60 μA to 180 μA, as shown by the vertical broken-line arrow in Fig. 2-1, there will be a change of collector current from 3.25 to 7.75 mA, as shown by the horizontal dotted arrows. The equation for calculating the current gain is

$$\beta = \frac{dI_c}{dI_b} \quad \text{(with } E_c \text{ constant)} \tag{2-1}$$

Equation (2-1) shows that the current gain is equal to the change of collector current dividend by a change of base current. When the values obtained from the chart in the foregoing example are set down, the equation shows that the current gain for this particular transistor circuit is 37.5, calculated as shown below:

$$\beta = \frac{dI_c}{dI_b} = \frac{7.75 \text{ milliamperes} - 3.25 \text{ milliamperes}}{180 \text{ microamperes} - 60 \text{ microamperes}}$$

$$= \frac{4.5 \text{ milliamperes}}{120 \text{ microamperes}} = \frac{0.0045}{0.000120} = 37.5$$

This calculation indicates that the signal current in the base emitter side is amplified 37.5 times in the collector side. The relation of such current gain to power gain is covered more fully later.

2-2 OTHER COMMON-ELEMENT BASIC CIRCUITS

In addition to the common emitter and common-source circuits shown in Figs. 1-7 and 1-9, two other basic types are used. As shown in Fig. 2-2, these comprise the common base and common collector for the bipolar transistors, and the common drain and common gate for JFET units. (Equivalent circuits for the MOSFET are given in Fig. 2-4.)

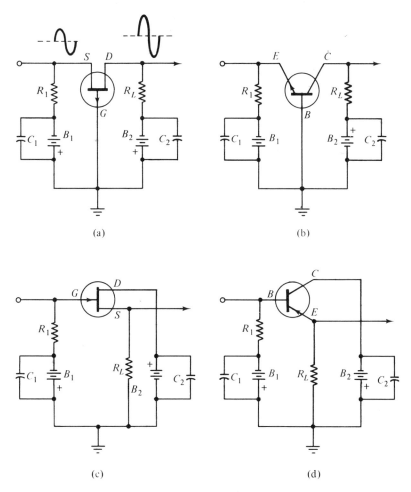

Figure 2-2 Common gate, base, drain, and collector circuits (basic)

The circuit shown in Fig. 2-2(a) uses a *p*-type JFET in a common-gate circuit; this compares to the transistor common-base shown in (b). In each of these circuits there is no phase reversal between the input signal and the amplified version appearing at the output. Because the gate in Fig. 2-2(a) and the base in (b) are connected to ground, effective isolation exists between the input and output circuits, and better performance and stability are obtained in some applications, as covered more fully in Chapter 3. (If a series resistor and bypass capacitor are used between the gate or the base and ground, these elements would still be at virtual ground for the signal, since the bypass capacitors provide a low reactance for the signals.)

For each of the circuits [in (a) and (b)] the input signal is impressed between ground and the input line. Batteries are bypassed to prevent signal energy from developing across the internal resistance.

If power supplies are used, the last filter capacitor of the supply could act as the necessary bypass to route signals around the supply. This, however, forms a lengthy ground return for the signals, and losses would occur. Thus, decoupling systems are extensively used to form short paths for signal energies (see Sec. 3-8).

When the positive alternation of the input sine-wave signal appears between the source and ground (gate) of the circuit in Fig. 2-2(a), the bias furnished by battery B_1 is opposed, and hence total bias is reduced. This also decreases the depletion region controlling the majority current carriers between the output circuit (drain and gate); thus more current flows through R_L. Since the voltage-drop polarity across R_L from B_2 is positive toward the drain element, a positive alternation of the output signal is produced. Hence, the output signal is in phase with the input signal as shown in Fig. 2-2(a).

For the *npn* circuit in Fig. 2-2(b), a positive signal alternation between the emitter and ground (base) also decreases the forward bias supplied by B_1, and such a bias reduction also decreases the current flow in the collector circuit. Consequently, the voltage drop across R_L decreases. Since, however, the polarity of battery B_2 places a negative polarity at the collector end of R_L, a reduction in negative voltage drop across R_L increases the signal level above ground (more positive), thus producing the same phase at the output as prevails for the input signal.

For the circuits shown in (a) and (b), negative alternations of the input signal would have the opposite effect on the forward bias; hence the output signal would follow in phase that of the input signal and a negative alternation would be produced. Similar characteristics prevail if the *p*-channel FET in (a) is replaced with an *n*-channel type, or if the *npn* transistor in (b) is replaced by a *pnp* type. In either instance, the polarities of B_1 and B_2 would have to be reversed.

For the circuit shown in Fig. 2-2(c), a common-drain system is used. (The drain element is at signal ground because of the bypass effect of capacitor C_2.) The output for the circuit in (c) is obtained from the load resistor (R_L) in the source circuit as shown. Again, there is no phase reversal between the input and output signals. Signal voltage gain is less than unity, although signal current gain may be achieved. This circuit is useful for impedance-matching purposes, such as for obtaining a low output impedance with a high input impedance. In logic systems the circuit also permits signal transfer without changing the logic 1 or 0 representation that would occur if the circuit had a signal phase reversal. Since the phase of the output signal "follows" that of the input signal, the circuit has been termed a *source follower*.

Figure 2-2(d) shows the transistorized version of the FET circuit in (c). The circuit in (d) has the collector at signal ground by virtue of the bypass effect of capacitor C_2, thus making the circuit a *common-collector* type. The output is procured from the load resistor R_L in series with the emitter as shown, thus forming an *emitter-follower* circuit. The signal phase at the output is again the same as the input. As with the circuits in (a) and (b), different types of FET and transistor units can be substituted [p-channel for (c) and *npn* for (d), for instance], provided that appropriate changes of battery polarity are made. Otherwise, operational characteristics are virtually the same, and in all instances the phase of the output signal "follows" that of the input signal. For the circuit in (d), as with the FET source follower, signal-voltage gain is less than unity, though signal-current gain can be obtained.

2-3 FET CHARACTERISTIC CURVES

Voltage and current relationships for the FET can also be utilized to obtain a set of characteristic curves, as was done for the transistor in Fig. 2-1. Typical sets of curves for the MOSFET are shown in Fig. 2-3. Here, the drain to source (V_{DS}) voltage is plotted along the x axis and the drain current (I_D) along the y axis. The various curves represent those obtained for a specific dc gate (V_{GS}) voltage (that is, the voltage between the gate and the source terminals).

The curves shown in Fig. 2-3(a) are for an enhancement FET, while those in (b) are for the depletion type. As with the curves shown in Fig. 2-1 for the bipolar transistor, specific information with regard to the potential operational characteristics can be obtained from a set of curves. Also, as shown later in this chapter, load lines can be drawn for information concerning the dynamic conditions which exist for signal voltages and currents. (See also Sec. 3-15.)

One important designation for FET characteristics is the pinch-off voltage (V_p or V_{po}). This is the point where the gate bias voltage V_{GS} is such that it causes the drain current to drop to zero for a specific value of V_{DS}. (See Secs. 2-5 and 3-14.)

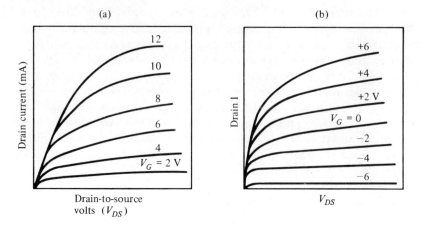

Figure 2-3 Curves for enhancement and depletion FETs

Another important FET characteristic is the *transconductance*, expressed as

$$g_m = \frac{dI_d}{dV_g} \quad \text{(with } V_{DS} \text{ constant)} \tag{2-2}$$

The unit value is the *mho*, which is *ohm* spelled backwards. The g_m is obtained by taking the ratio of a small change in drain current (dI_d) to a small change in gate voltage (dV_g) from a set of characteristic curves. Thus, g_m is a *figure of merit* concerning the ability of the gate voltage to control the drain current.

The transconductance for an FET is usually given with zero gate bias, and a minimum prevails at the pinch-off voltage. If the load resistance value is known, signal gain is found by

$$\text{Gain} = g_m \times R_L \tag{2-3}$$

2-4 SOURCE, GATE, AND DUAL-GATE

The common-source circuits for the FETs were shown in Figs. 1-9 and 1-10 and were compared to the common-emitter type for the bipolar transistor. Comparable MOSFET circuits for those shown in Fig. 2-2 are given in Fig. 2-4. In (a) the depletion-type *source-follower* circuit is shown; as with the emitter follower, the output is obtained from an unbiased resistor, as shown. There is no phase reversal between the input signal and that obtained from the source-follower output. An

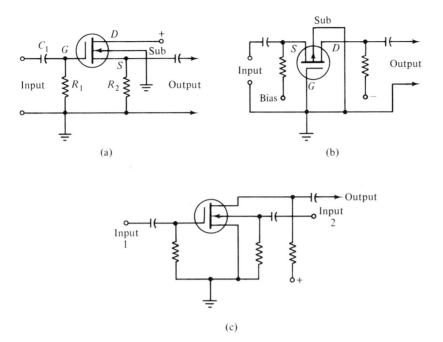

Figure 2-4 Source follower, common gate, and dual-gate circuits (basic)

n-type FET is shown, though a p-channel can, of course, also be used with appropriate polarity changes of the applied potentials.

A p-channel depletion-type MOSFET circuit is shown in Fig. 2-4(b); it is the grounded-gate type, comparable to the grounded base for the bipolar transistor. Again, there is no phase reversal between input and output signals. A dual-gate n-channel is shown in (c), with inputs available at the gate and sub terminals as shown. This is a grounded-source circuit. (See also Secs. 2-9 and 3-15.)

2-5 TRANSFER CHARACTERISTICS

A common term used in relation to junction transistors and FET units is the *signal-transfer ratio*. For the FET, the g_m can be considered as the signal-transfer ratio. The forward current-transmitter ratio for the grounded-emitter circuit has already been given in Eq. (2-1). For the grounded-base circuit shown in Fig. 2-2(b), the ratio is termed *alpha* (α) and it is the ratio of a change of collector current I_c to a change of emitter current I_e:

$$\alpha = \frac{dI_c}{dI_e} \quad \text{(with } E_c \text{ constant)} \tag{2-4}$$

A set of transfer-characteristic curves for the bipolar transistor is shown in Fig. 2-5. These curves were obtained for several base-current values plotted against collector current and base-to-emitter voltage. The transfer characteristics shown in Fig. 2-6 relate to the three operational modes for the field-effect transistor. (See Sec. 1-9 for a discussion of the operational modes and relative bias factors for the JFET and MOSFET units.) The curve designated as A represents the depletion mode, curve B indicates the depletion-enhancement mode, and curve C shows the enhancement mode. Curve A can represent both the JFET and the MOSFET (IGFET), whereas B and C are for the MOSFET type only.

The point designated as V_p is the pinch-off voltage discussed in Sec. 2-3, and the I_{DSS} point is where the drain current reaches the $V_G = 0$ curve, as more fully discussed in Sec. 3-14 and illustrated in Fig. 3-12. As shown for the A curve, the slope of the tangent line is marked g_m and represents the transconductance at that point. Hence, such a tangent line can be drawn on any part of the slope to represent the g_m value at that particular point.

2-6 NETWORK PARAMETERS

Operational characteristics (parameters) of transistors can be obtained by setting up circuit networks which are the equivalent of the

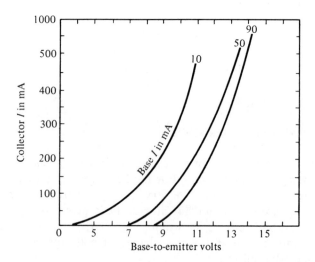

Figure 2-5 Typical transistor transfer characteristics

Network Parameters / 29

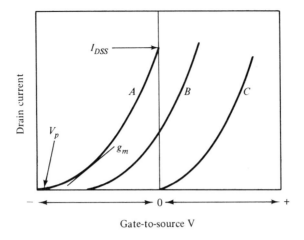

Figure 2-6 Transfer characteristics of FET (three modes)

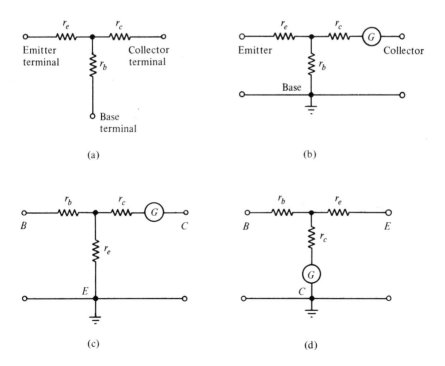

Figure 2-7 Transistor network equivalents

original circuits but which lend themselves more readily to the specific collection of data. Several types of parameters are in general use, and a familiarization with their identity and application provides for an in-depth understanding of transistor dynamic operating conditions and circuit configurations.

A good introduction to transistor parameter applications consists in initially evaluating passive networks. Actually, a transistor can be considered as a resistive device represented as shown in Fig. 2-7(a). Here the emitter is denoted by a resistor R_e, the base by r_b, and the collector by R_c. Together these units form a T network of resistors with ohmic values as would be obtained by taking dc measurements. Since such a T network is a *passive* type, it is not a true representation of a transistor, because such a three-resistor combination does not have amplifying characteristics. Also, the network shown in (a) is a three-terminal device, whereas a common input-output lead is present in a transistor circuit; hence, for practical purposes, the transistor should be represented by a *four-terminal* network.

The four-terminal equivalent circuit is shown in Fig. 2-7(b) and represents the common-base (grounded-base) circuit shown earlier in Fig. 2-4(b). Instead of a passive network, a generator (G) is shown in the collector lead, thus establishing this circuit as an *active* network. Hence, the amplifying function of the transistor is represented by the generator in the output. A grounded-emitter representation is shown in Fig. 2-7(c). For the grounded emitter, the generator is placed in the output line to represent the emitter-collector section. Now, the left arm of the T network is shown as r_b for the equivalent base resistance that now forms the input terminal, and r_e as the representative grounded-emitter resistance. The active network for the grounded collector is shown in Fig. 2-7(d).

At low signal frequencies (or with dc), the active resistance network is a close approximation of the transistor circuit's characteristics. At high signal frequencies, however, it becomes an impedance network, because the internal capacitances, with their decreasing reactances, are influencing factors. Input and output impedances are also affected by the ohmic value of the load resistance applied to the output, as well as of the resistance of the circuit or device applied to the input. Hence, for a true evaluation of circuit parameters, the internal network values must be taken into consideration in conjunction with the external networks that are applied, so that maximum signal-power transfer and maximum circuit efficiency are obtained. In the discussions that follow, we shall be concerned primarily with the resistive characteristics of the equivalent networks for a clearer understanding of the notation employed and the methods utilized.

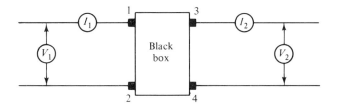

Figure 2-8 "Black box" representation

In the mathematical and logical analysis of networks it is often expedient to consider the network as being contained in a black box wherein we have internal components of unknown values. This concept has been widely used by engineers in design practices and is a useful procedure. To analyze the internal circuit, we read voltages and currents, or we apply test signals to the input and output terminals and evaluate their effects. The voltage and current measurements are also undertaken during the open-circuit or short-circuit condition of either the input or output terminals.

Thus, the four-terminal equivalent network of a triode transistor can be represented as a black box, as shown in Fig. 2-8. Here, V_1 is the input voltage, V_2 the output voltage, I_1 the input current, and I_2 the output current. The resistance parameters of the transistor so represented have been symbolized by the so-called R parameters as well as by the more widely used h parameters. Initially we shall consider the R-type parameters, since they serve as a more solid foundation for understanding the h (hybrid) types.

Because $E/I = R$, the input resistance is notated as R_{11}, to identify that this R value is obtained by using the V_1 and I_1 values. Thus the first 1 of the subscript indicates V_1 voltage and the second 1 of the subscript the I_1 current. Thus, R_{11} is a measure of the resistance with the test voltage V_1 applied to the input terminals 1 and 2, with the output terminals 3 and 4 open; that is, $I_2 = 0$. Similarly, R_{21} indicates the forward transfer resistance, with the test voltage V_1 applied to the input terminals again, and the ratio of V_2/I_1 taken for the R_{21} value. When the test voltage is applied to the output terminals 3 and 4 and the input terminals are left open ($I_1 = 0$), we obtain

$$R_{12} = \frac{V_1}{I_2} = \text{reverse transfer resistance (feedback resistance)}$$

$$R_{22} = \frac{V_2}{I_2} = \text{output resistance}$$

From the foregoing we can derive the loop equations for the transistor network.

$$V_1 = R_{11}I_1 + R_{12}I_2 \tag{2-5}$$

$$V_2 = R_{21}I_1 + R_{22}I_2 \tag{2-6}$$

The amplifying ability of the transistor network relates to the mutual resistance (r_m), and current amplification is referred to as *alpha* (α). Related to the R parameters, these are:

$$r_m = R_{21} - R_{12} \tag{2-7}$$

$$\text{alpha} = \frac{R_{21}}{R_{22}} \tag{2-8}$$

The *r*-value relationships of Fig. 2-7(a) (common base) become

$$r_e = R_{11} - R_{12} \tag{2-9}$$

$$r_c = R_{22} - R_{21} \tag{2-10}$$

$$r_b = R_{12} \tag{2-11}$$

The equivalent generator voltage is equal to $I_1 (R_{21} - R_{12})$. The resistive parameters can also be utilized for the dynamic characteristics of the network under signal conditions. Using d to indicate a quantity change, and holding i_c constant, we get

R_{11} = slope of curve dV_e/dI_e

R_{21} = slope of curve dV_c/dI_e

Holding i_e constant produces the following:

R_{12} = slope of curve dV_e/dI_c

R_{22} = slope of curve dV_c/dI_c

When audio, RF, and similar ac-type signals are involved, both the *reactive* as well as the *resistive* components become a factor. Consequently, the R designations are replaced by the impedance Z. Now the black-box representation of the electronic circuit input would have an impedance and, with an open circuit at the output ($I_2 = 0$), we obtain

$$Z_{11} = \frac{V_1}{I_1} \quad (I_2 = 0) \tag{2-12}$$

Again with an open circuit at the output, we obtain the forward transfer impedance comparable to R_{21}:

$$Z_{21} = \frac{V_2}{I_1} \quad (I_2 = 0) \tag{2-13}$$

The reverse transfer impedance is procured with the input circuit open ($I_1 = 0$) and this produces

$$Z_{12} = \frac{V_1}{I_2} \quad (I_1 = 0) \tag{2-14}$$

The output impedance is also obtained with the input held in the open-circuit condition:

$$Z_{22} = \frac{V_2}{I_1} \quad (I_1 = 0) \tag{2-15}$$

Now we can rewrite the loop equations given earlier [Eq. (2-5) and (2-6)] as follows:

$$V_1 = Z_{11}I_1 + Z_{12}I_2 \tag{2-16}$$
$$V_2 = Z_{21}I_1 + Z_{22}I_2 \tag{2-17}$$

2-7 HYBRID (h) PARAMETERS

Open-circuit conditions were used to obtain the R parameters, and these are representative of *constant-voltage* types. Often, however, it is more convenient to use the *constant-current* analysis, shorting out terminals as needed. By combining the constant-voltage and constant-current procedures, a more desirable type of parameter is obtained—one that has been used generally by engineers and industry. The combination has led to the term *hybrid* or *h parameter,* which refers to the type primarily used for obtaining operational characteristics of the bipolar transistors. The *Y*-type parameter is also useful, and applications to the unipolar transistor (FET) are shown later.

For the h parameters, the notations that follow again refer to the black-box concept illustrated in Fig. 2-8:

$$h_{11} = \frac{V_1}{I_1} \quad \text{(an input \textit{impedance} parameter, with output terminals 3 and 4 shorted and } V_2 = 0\text{)}$$

$h_{12} = \dfrac{V_1}{V_2}$ (a reverse-transfer *voltage ratio*, with input terminals 1 and 2 open and $I_1 = 0$)

$h_{21} = \dfrac{I_2}{I_1}$ (a forward-transfer *current ratio* with output terminals 3 and 4 shorted and $V_2 = 0$)

$h_{22} = \dfrac{I_2}{V_2}$ (an output *admittance* term with input terminals 1 and 2 open and $I_1 = 0$)

Basic calculations can be used if necessary to convert R to h or h to R. Parameter R_{11}, for instance, is equal to $(h_{11}h_{22} - h_{12}h_{21})/h_{22}$. Similarly, $R_{12} = h_{12}/h_{22}$ and $R_{21} = h_{21}/h_{22}$. The equations for the R and Z parameters [Eqs. (2-5), (2-6), (2-16), (2-17)] involve input voltage and output voltage (V_1 and V_2). For the h parameters, however, the equations relate to input voltage and output current, as follows:

$$V_1 = H_{11}I_1 + H_{12}V_2 \qquad (2\text{-}18)$$

$$I_2 = H_{21}I_1 + H_{22}V_2 \qquad (2\text{-}19)$$

Standards have been adopted for letter subscripts for easier identification of the h parameters. The first subscript designates the characteristic, i for input, o for output, f for forward transfer, and r for reverse transfer. The second subscript designates the circuit configuration, with b for common base, e for common emitter, and c for common collector. Thus, h_{11} can be indicated as h_{ib} for the input h of a common base network. Similarly, h_{12} can be written as h_{rb} for reverse transfer in the common base network. This avoids the confusion that might result by simply using h_{11}, for instance, without designating whether it is in reference to common base, common emitter, or common collector.

Figure 2-9(a) shows a basic grounded-emitter circuit. Applied to the input is the signal voltage to be amplified. This could be symbolized as e_s for *signal voltage* or as e_g for *generator voltage,* as shown. The resistance associated with the input signal is marked R_g. The load resistance applied to the output is indicated by the standard R_L symbol. The dc power sources E_1 and E_2 are effectively bypassed by capacitors C_1 and C_2; hence they are not part of the active parameter analysis.

The equivalent circuit for the grounded emitter is shown at (b), with appropriate hybrid symbols with letter subscripts. The base current is indicated as i_b and collector current as i_c. The input voltage e_g also develops across R_g; hence the actual voltage applied to the base and emitter is indicated as e_i. The output voltage across the load resistor is

Hybrid (h) Parameters / 35

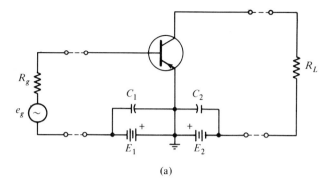

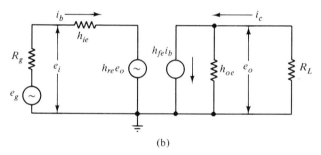

Figure 2-9 Grounded-emitter circuit and equivalent

shown as e_o. The emitter input resistance is given as h_{ie}, and the output admittance or conductance is given as h_{oe}, measured in mhos. Note the use of the reverse-voltage transfer ratio symbol h_{re} for the emitter circuit. (In a common base this would be h_{rb}.) The forward current ratio is designated as h_{fe}, indicating the emitter designation.

The forward current-transfer ratio h_{fe} is now i_c/i_b (with $e_o = 0$), which is the same as Eq. (2-1), given earlier for signal current gain β (beta). Manufacturers have used A_i for signal current gain, symbolizing current amplification. The full equation is

$$A_i = \frac{i_c}{i_b} = \frac{h_{fe}}{1 + h_{oe}R_L} \tag{2-20}$$

While this equation refers to the common-emitter circuit of Fig. 2-9, it also applies to the grounded-base or grounded-collector circuits. The equation remains the same except for a change from e to b or c for the second subscript to suit the circuit configuration. Other equations used to analyze circuit parameters are also applicable to the three basic circuit systems of transistors, the appropriate second subscript being used to denote emitter, base, or collector. The following example shows the application of Eq. (2-20):

Example: A transistor has a low signal designation of $E_1 = 5$ V and $I_b = 0.1$ mA. The manufacturer's h ratings are

$$h_{ie} = 2800 \ \Omega$$
$$h_{oe} = 43 \ \mu\text{mhos}$$
$$h_{fe} = 110$$
$$h_{re} = 7.5 \times 10^{-4}$$

Assume that a load resistance of 10,000 Ω is to be used. What is the current gain?

Solution: From Eq. (2-20):

$$A_i = \frac{110}{1 + (43 \times 10^{-6} \times 10,000)} = 77$$

The input resistance R_i can be found by the following equation:

$$R_i = h_{ie} - \frac{h_{fe} h_{re} R_L}{1 + h_{oe} R_L} \qquad (2\text{-}21)$$

Using the h values given above for the transistor, with a load resistance of 10,000 Ω, the following shows the application of Eq. 2-21:

$$R_i = 2800 - \frac{110 \times 7.5 \times 10^{-4} \times 10,000}{1 + (43 \times 10^{-6} \times 10,000)} = 2223 \ \Omega$$

The single-voltage-gain amplification of the transistor circuit is found by

$$A_e = \frac{e_o}{e_i} = \frac{1}{[h_{re} - (h_{ie}/R_L)][(1 + h_{oe} R_L/h_{fe})]} \qquad (2\text{-}22)$$

Again, using the values given for the transistor as an example, we find the voltage amplification:

$$A_e = \frac{1}{[0.00075 - (2800/10,000)] \cdot [1 + (43 \times 10^{-6} \times 10,000)/110]} = 277$$

The power gain of the transistor circuit is found by multiplying the signal-current gain A_i by the signal-voltage gain A_e:

$$A_p = A_e A_i \qquad (2\text{-}23)$$

Applying the two values previously obtained for the transistor used as an example, we find the power gain to be

$$A_p = 277 \times 77 = 21{,}329$$

2-8 Y PARAMETERS

As mentioned earlier, the admittance (Y) parameters are most useful for investigating the operational characteristics of the field-effect transistor. Since $Y = 1/Z$, the following replace the h notations given earlier for the black box of Fig. 2-8:

$Y_{11} = \dfrac{I_1}{V_1}$ (an input *impedance* parameter with output terminals 3 and 4 shorted and $V_2 = 0$)

$Y_{12} = \dfrac{I_1}{V_2}$ (reverse transfer admittance with input terminals 1 and 2 shorted and $V_1 = 0$)

$Y_{12} = \dfrac{I_2}{V_1}$ (forward transfer admittance with output shorted and $V_2 = 0$)

$Y_{22} = \dfrac{I_2}{V_2}$ (output admittance with input shorted and $V_1 = 0$)

Instead of the input voltage and output current h equations [Eqs. (2-18) and (2-19)], the FET equations for Y parameters involve input current I_1 and output current I_2:

$$I_1 = Y_{11} V_1 + Y_{12} V_2 \qquad (2\text{-}24)$$
$$I_2 = Y_{21} V_1 + Y_{22} V_2 \qquad (2\text{-}25)$$

As with the letter subscripts used for easier identification of the h parameters, we can use i, o, f, and r for FET units to describe parameters of input, output, forward transfer, and reverse transfer. A second subscript is used to identify gate, source, or drain. Thus, the I_1 and I_2 equations [Eqs. (2-24) and (2-25)] for *common-source* circuitry become

$$I_g = Y_{is}V_g + Y_{rs}V_d \tag{2-26}$$

$$I_d = Y_{fs}V_g + Y_{os}V_d \tag{2-27}$$

For common-gate or source-follower design, the second subscript of Eqs. (2-26) and (2-27) are changed accordingly. Thus, for common-gate circuitry,

$$I_s = Y_{ig}V_s + Y_{rg}V_d \tag{2-28}$$

$$I_d = Y_{fg}V_s + Y_{og}V_d \tag{2-29}$$

For comparison purposes the following shows the related equations for the Z, h, and Y parameters for the common-emitter circuit. Symbols thus relate to the transistor, where i_c is collector current. The symbol i_o could also be used as a general indication of output current. Similarly, V_o indicates output voltage, though V_c could be used to designate collector voltage. For FET, appropriate symbols designate drain, source, and gate parameters as shown in Eqs. 2-26 through 2-29.

$$Z \begin{cases} V_i = Z_{ie}i_b + Z_{re}i_c \\ V_o = Z_{fe}i_b + Z_{oe}i_c \end{cases} \tag{2-30}$$

$$h \begin{cases} V_i = h_{ie}i_b + h_{re}V_o \\ i_c = h_{fe}i_b + h_{oe}V_o \end{cases} \tag{2-31}$$

$$Y \begin{cases} i_b = y_{ie}V_i + y_{re}V_o \\ i_c = y_{fe}V_i + y_{oe}V_o \end{cases} \tag{2-32}$$

2-9 MOSFET GAIN FACTORS

The common-source FET circuit shown in Fig. 1-10 has a high input impedance (in the megaohms) and an output Z in the 100-kΩ range. The current-gain equation [Eq. (2-1)] used for the grounded-emitter bipolar transistor does not apply for the FET units, because the high input impedance relates signal *voltage* variations to signal *current* changes between drain and source. Thus the transconductance (g_m) given in Eq. (2-2) is appropriate for the FET units to describe the characteristics. We can, however, relate the g_m to other values and obtain an equation that would indicate the voltage gain (without feedback) for the common-source MOSFET circuit:

$$A_e = \frac{g_m r_d R_L}{r_d + R_L} \tag{2-33}$$

where g_m is the gate-to-drain transconductance
r_d is the drain resistance
R_L is the effective load resistance at the output

As discussed in Sec. 1-11, the use of an unbypassed source-to-ground resistor [such as R_2 in Fig. 1-10(c)] causes degeneration. Because the signals that develop across such a resistor vary inversely to the signals being amplified, a form of degenerative feedback is developed. (See also Sec. 6-14, "Operational Amplifiers," for other feedback factors in amplifiers.) With the source resistor unbypassed, the common-source voltage gain is expressed as A' [see Eq. (6-9)] and in equation form is

$$A' = \frac{g_m r_d R_L}{r_d + (g_m r_d + 1)R_s + R_L} \tag{2-34}$$

where R_s is the ohmic value of the unbypassed source resistor.

When the source resistor is not bypassed, the output impedance Z_o is increased because of the series-resistance effect of the unbypassed resistor with the output impedance. Thus, the new value of Z_o is

$$Z_o = r_d + (g_m r_d + 1)R_s \tag{2-35}$$

For the source-follower circuit shown in Fig. 2-4(a), the source resistor cannot be bypassed, because the signal is taken from this point and would be lost by shunt if bypassed. Thus, the common-drain (source follower) circuit is automatically an inverse-feedback system with degeneration. Thus, the gain is

$$A' = \frac{g_m R_s}{1 + g_m R_s} \tag{2-36}$$

For the common-gate circuit shown in Fig. 2-4(b) the voltage gain can be found by using the following equation:

$$A_e = \frac{(g_m r_d + 1)R_L}{(g_m r_d + 1)R_a + r_d + R_L} \tag{2-37}$$

where R_a is the resistance applied to the input by the signal-source circuitry.

Where r_d has a sufficiently high value compared to R_L to minimize shunting effects, a more general form of Eq. (2-37) can be used:

$$A_e = \frac{g_m R_L}{g_m + 1} \tag{2-38}$$

2-10 INPUT-OUTPUT R EQUATIONS

For the source-follower circuit using the MOSFET, the input resistance R_i is approximately equal to the gate-to-ground resistance R_1 in Fig. 2-4(a). (See also Secs. 3-14 and 3-15 for related equations and discussions.) When, however, resistor R_1 is returned to the source terminal of the FET, the input resistance changes and is now calculated by the following equation:

$$R_i = \frac{R_1}{(1 - A')} \tag{2-39}$$

where A' = the signal-voltage amplification with feedback, as determined by Eq. (2-36).

Thus, if $R_1 = 200 \text{ k}\Omega$ and A' is 0.3, we have the following result:

$$R_i = \frac{200{,}000}{(1 - A')} = \frac{200{,}000}{0.7} = 285{,}714 \ \Omega$$

The output resistance of the source follower is found by

$$R_o = \frac{r_d R_s}{(g_m r_d + 1)R_s + r_d} \tag{2-40}$$

For the common-gate circuit shown in Fig. 2-4(b) the input impedance can be ascertained by

$$Z_i = \frac{g_m R_s + 1}{g_m} \tag{2-41}$$

For the ohmic value of the output impedance Z_o we relate the drain-to-gate resistance to the load resistance:

$$Z_o = \frac{r_d R_L}{r_d + R_L} \tag{2-42}$$

2-11 TERMINAL IDENTIFICATION

Terminals or lead connections for a given transistor may vary to a considerable extent among manufacturers; hence no set rules can be

Terminal Identification / 41

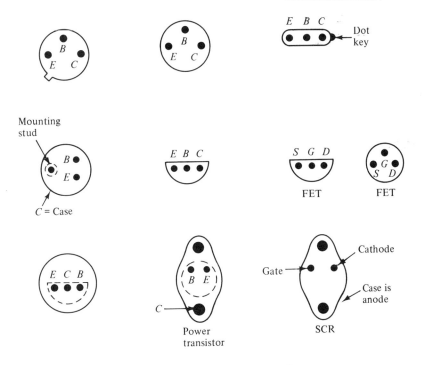

Figure 2-10 Solid-state terminal variations

used as a guide to identification of leads from just the diagram representation. Some general types are illustrated in Fig. 2-10 for reference purposes and as an indication of the styles and types usually encountered. All are bottom views of the transistor, both for the junction and the FET types. Note the emitter-base-collector sequence for the first three shown in Fig. 2-10. The leads are arranged in a triangular fashion so that identification can readily be made. On occasion, however, a metal extension may show the emitter location, and sometimes a dot key or red stripe may indicate and identify the proper sequence. For the higher power transistors, a mounting stud is provided, or holes are given for bolting the unit to the chassis.

Although the lead sequences shown in Fig. 2-10 have been widely used, many junction transistors and FET units with the same lead spacings may have other sequences. Consequently, a transistor reference manual must be consulted for positive identification of a transistor's number.

The identifying (numbering) systems used for transistors also vary considerably among manufacturers, even though the characteristics of the transistors are similar. Transistor numbering systems usually include letters as well as numbers, and hence appear as 2N212, 2SC984,

2N1069, and 2N5296. Again, reference should be made to the manufacturer's specifications or transistor manuals for an identification of the characteristics of a particular type. Manuals are also available that list transistors having similar characteristics but different numbers.

2-12 HEAT-SINK FACTORS

Transistors with wattage ratings above one or two watts begin to generate sufficient heat that special precautions must be observed to dissipate such heat. With the larger power transistor, the heat dissipation devices (heat sinks) are essential; otherwise, the transistor will burn out in a short period of time. Typical higher power transistors are illustrated in Fig. 2-11. The type shown in (a) provides a flat metal section, which is bolted to the chassis for heat dissipation purposes. Since the collector is electrically connected to the metal shell, an insulating mica washer must usually be used between the transistor and the chassis to prevent a short circuit. The type in (b) is provided with a bolt at the center so that it can be fastened to the chassis. Again, the chassis

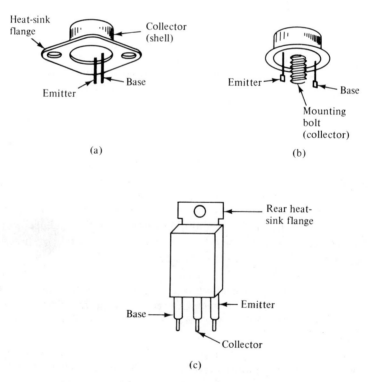

Figure 2-11 Various power transistors

(metal) acts to dissipate the heat generated by the transistor. Still another type is that shown in (c), where a rear heat-sink flange is an integral part of the transistor. A mounting hole is provided so that the heat sink can be bolted to a chassis section for heat dissipation purposes. (Lead designations shown may vary for different type transistors.)

Several heat-sink methods are illustrated in Fig. 2-12. In (a) is shown the method used for insulating the collector shell from the metal chassis. A thin mica insulating gasket is used as shown; this is preperforated to accommodate the mounting bolts and the two prongs representing the emitter and base terminals. A special silicon-base thermal-conductive grease is applied to both sides of the gasket to ensure maximum heat transfer to the chassis. Thus, in such an insulation, the metal chassis acts as the heat sink by absorbing the excessive heat of the transistor. The use of the silicon-base grease is essential for maximum heat conduction to the chassis.

For plastic chassis (such as the preconstructed type) a separate heat-dissipating flange is used as a heat sink, as shown in Fig. 2-12(b). This flange is usually made of aluminum and has sufficient surface area for adequate dissipation of the heat generated by the transistor during normal usage. As with the transistor installation shown in (a), however, it is still necessary to use the silicon-base thermal-conduction grease. If

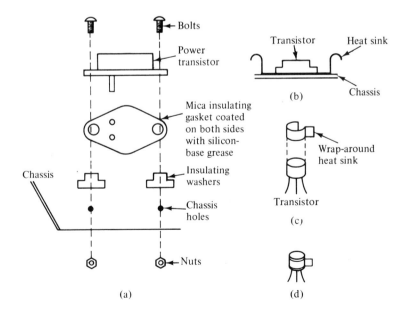

Figure 2-12 Heat sinks

44 / Transistor Characteristics and Parameter Factors

the heat sink is not in contact with any ground connections, the mica gasket is dispensed with.

For the smaller transistors, a wraparound metal heat sink is often employed, as shown in Fig. 2-12(c). The heat sink has a small extension, which is sufficient for dissipating the heat when the heat sink is pressed over the transistor casing, as shown in (d).

2-13 TRANSISTOR TESTING

Many engineers and technicians routinely check new transistors before installing them into an experimental project or when replacing a defective one. Although quality control in transistor manufacture may be high, on occasion a defective transistor or one having characteristics below standard may be encountered. Usage of such an off-value unit will provide poor operation and may result in fruitless testing of associated circuit components.·

There are many types of transistor testers on the market, and the number of features incorporated in a particular unit usually depends on the cost. The more inexpensive types simply test for leakage between elements and test for shorted conditions between emitter and base, or between collector and emitter. For the higher quality commercial transistor testers, a beta range is also provided, as shown in Fig. 2-13. For this unit, the beta gain is represented in micromhos.

Dual scales are sometimes provided in some instruments with ranges of up to 500 micromhos for one scale, and another giving still higher ranges for greater versatility in testing a variety of solid-state devices.

A leakage test is useful for determining mA or μA of current leakage (usually between the base and collector junction under reverse-current conditions). This is an important test, since even one or two μA

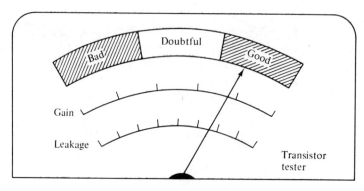

Figure 2-13 Transistor tester dial

of leakage can cause poor performance in many transistors. Such test instruments also provide for leakage checks for field-effect transistors between gate and source elements. Some commercial solid-state test instruments provide for in-circuit checks, eliminating the need for unsoldering leads when transistors are tested that are suspected of being defective. Sometimes testers also have provisions for identifying emitter, base, and collector leads in a transistor where these leads are not identified, and for ascertaining whether the transistor is a *pnp* or an *npn* type.

CHAPTER 3

Amplifier Circuits

3-1 LOSS FACTORS

The purpose of an amplifier is not confined solely to the production of a signal having a greater amplitude at the output than was applied to the input. Of concern also is the need for maintaining a flat frequency response throughout the amplifier, so that no signals of particular frequencies will be attenuated or amplified to a greater degree than other signals. To keep loss factors at a minimum, components and conditions that contribute to such signal attenuation must be recognized and steps taken to minimize their effect. (Special amplifiers are covered in Chapter 6.)

When a transformer is used to couple one amplifier stage to another, both the primary and secondary windings contribute to an uneven frequency response. At the higher audio frequency signals, the reactances of the transformer inductors are high, and, consequently, a higher order of signal voltages occur across the transformer. At low signal frequencies, however, the transformer reactances are also low, and the low-frequency signals are not amplified to the same degree as the higher frequency signals. Also, the transformer inductors have distributed capacitances, which occur between individual turns of wire as well as between layers of the wire turns making up the primary and secondary coils. Thus, higher frequency signals cause the distributed capacitances to have low reactances, and the latter tend to shunt signals at the higher frequencies. The result of this is a loss of low-frequency signals because of inductive reactances in the transformer and a loss of high-frequency signals due to the distributed capacitances. This is known as *frequency distortion*. To minimize such losses, it becomes necessary to select transformers with larger cores and to choose core metals that increase permeability. The increase in the permeability results in an increase in inductance, permitting fewer turns of wire to produce the required inductance and thus also lessening the total distributed capacitance. (See also Sec. 3-4.)

When a capacitor is used for interstage coupling, loss problems also occur. Signals with higher frequencies find a low reactance in the coupling capacitor and thus encounter little opposition in reaching the next amplifier stage input. Signals with lower frequencies, however, encounter a higher series reactance and thus are diminished in transferring across the coupling capacitor. If the latter's capacitance value is increased in an effort to accommodate lower frequency signals, the reactances for lower frequency signals will decrease, but the larger physical size of the capacitor may set up shunt capacitances to a metal chassis (or nearby components) and may thus shunt some of the higher frequency signals. The stray capacitance from the body of a capacitor to the chassis may be small, but the higher frequency signals may encounter a sufficiently low reactance that an appreciable portion of the high-frequency signal energy is lost. Even undue lengths of wiring or printed-circuit conductors that connect a coupling capacitor from the output circuit of one stage to the input circuit of the following stage may introduce losses. Long leads tend to increase inductive effects, because even a short length of wire has some inductive characteristics. The inductance of the wire length may be small, but high-frequency signals encounter a greater inductive reactance in the wire length than low-frequency signals; hence the high-frequency signals suffer some attenuation. This factor is, of course, more prevalent at RF regions. Also, any wire connecting circuits may present some shunt capacitance to a metal chassis or other nearby components. If such wires carry signals, the stray capacitance will provide a shunt reactance that will contribute to signal losses.

In the design of amplifiers, loss factors may of necessity be tolerated in order to keep construction costs at a minimum. The amount of signal loss that is to be tolerated is directly related to the particular market level for which the product containing the amplifier is aimed.

Signal losses also occur because of the interelement capacitances of solid-state devices. Such capacitances exist between base and emitter, between collector and base, and between collector and emitter. For a minimum of signal loss at high-frequency operation, transistors must be selected that are specifically designed for low shunt capacitances at the upper frequency ranges.

Loss factors that can contribute to poor performance are illustrated in Fig. 3-1, which summarizes the factors discussed in this section. In a well-designed audio amplifier, where all necessary signal-loss precautions have been observed, it is possible to obtain a substantially flat response for signals ranging in frequencies from as low as 30 Hz to well over the hearing limit of 20 kHz.

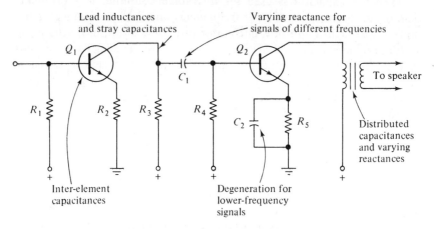

Figure 3-1 Amplifier loss factors

The overall gain that can be realized from an amplifier is dependent on the gain factor of the transistor, as well as on the value of the load resistor used and the output impedance of the transistor. As the signal frequencies are increased, upper limits will be reached in terms of signal transfer efficiency. For common-base circuits, the cutoff point for signal frequency is considered to be reached when the alpha value drops to 0.707 of the value obtained at 1 kHz. For the common-emitter circuit, the frequency cutoff is reached when beta drops to 0.707 of the 1-kHz value.

3-2 LOAD LINE FACTORS

Characteristic curves such as were shown in Fig. 2-5 display the amount of output current flowing through the transistor for a given bias and collector voltage. Such curves, however, do not indicate the dynamic characteristics, that is, the characteristics established when a signal is applied to the input of a circuit. During the presence of a signal, output current varies by an amount established by the amplitude of the input signal and the signal gain of the transistor. To illustrate the conditions that occur when a transistor operates with a signal input and is amplifying, a *load line* must be drawn to represent the dynamic characteristics. Such a load line is represented by a diagonal line drawn through the characteristic curves, as shown in Fig. 3-2. With a load line drawn, the dynamic operating conditions of the transistor can be calculated, and such information can be obtained as power output, percentage of distortion, and voltage amplitudes under operating conditions. Note the dashed-line curve in Fig. 3-2. This curve represents the constant power-dissipation line in watts for the collector side, as specified for a particular transistor. Thus, for the transistor graph in Fig. 3-2, the 60 mW is the maximum energy dissipation possible before an overload condition prevails. This dissipation curve is plotted along points established by the products I_c and V_c. Thus, at any point where the dashed curve intersects with abscissas and ordinates, the current-voltage product equals 60 mW for this particular transistor.

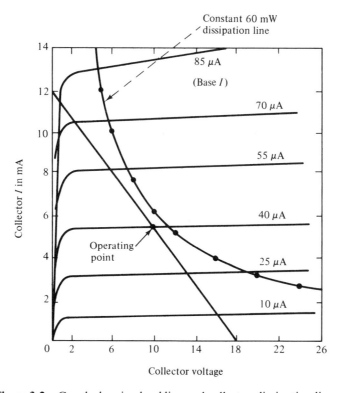

Figure 3-2 Graph showing load line and collector-dissipation line

Note that the load line is drawn in the area below and to the left of the dissipation line so that the given rating is not exceeded. Maximum power gain of the transistor is obtained when the load line is drawn tangent to the dissipation line. The load impedance (or resistance) may be calculated by the following equation, which is actually a computation of the *slope* of the load line:

$$R_L = \frac{(E_{max} - E_{min})}{(I_{max} - I_{min})} \tag{3-1}$$

This equation may also be expressed as

$$R_L = \frac{dV_c}{dI_c} \tag{3-2}$$

The operating point has been selected for class A operation, where the signal swing is on the linear portion of the curve above and below the operating point. This point is at 10 V for the voltage change, and 5.5 mA for the current change. The base current point is 40 μA, as shown in Fig. 3-2. At the operating point, $I_c V_c$ is .0055 × 10, which equals 55 mW, and hence is below the maximum permitted as indicated by the power dissipation line.

For the load resistance using the above equation, $dV = 0$ to 18, and $dI = 0$ to 12 mA; hence

$$\frac{18}{0.012} = 1500 \text{ ohms}$$

As the signal swings along the load line, it will not encroach on the base-current curvatures at the left, if class A operation is maintained.

The operating point as well as the load line slope can be altered to provide for a different value of load resistance or greater signal swing, etc. In Fig. 3-3, for instance, two load lines are displayed for comparison purposes. Both can use the same operating point, though each could have a different operating point if desired. In design practices, a load line is selected that provides a minimum of distortion for the power output required. The load line resistance values are again found by using Eq. (3-1) or Eq. (3-2). The power output and percentage of distortion obtained for each load line are determined as detailed in Sec. 3-3 which follows.

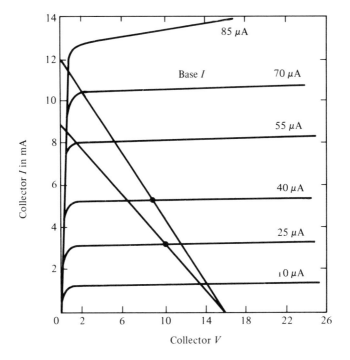

Figure 3-3 Load line comparisons

3-3 POWER AMPLIFIER LOAD LINE FACTORS

When a resistor is used as the load in an amplifier stage, the supply voltage indication on a load line drawing is the *right-hand termination* of the load line at the zero collector-current point. For the characteristics curves shown in Fig. 3-4, the supply voltage will be -28 V if this is a small-signal amplifier with a load resistor and coupling capacitor to the next stage. The signal-voltage swing across the load resistor for a signal input can then reach the maximum value of -28 V, since this is the highest voltage furnished by the power supply. The actual operating voltage would be only -12 V, because this is the voltage drop across the load resistance without a signal input.

When a transformer is used to couple the output of a transistor to a speaker or other transducer in power-amplifier circuitry, the actual supply voltage would only be -12 V for this particular transistor, the same potential as the operating voltage for the dynamic load line. The function, insofar as the dynamic characteristic is concerned with signal input, however, is essentially the same as when an actual load resistor in

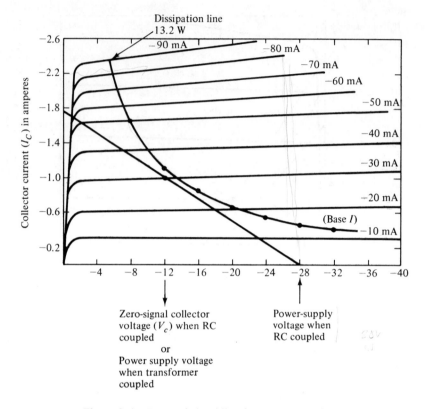

Figure 3-4 Dynamic load line for power transistor

series with the power source is present. With transformer coupling, the actual load is that which is applied to the secondary of the transformer, such as a loudspeaker voice coil. The function, in terms of collector-current swing, occurs because of the action of the inductance through which the collector current of the power amplifier must flow.

Thus, a swing of the input signal in a positive direction (a decrease in forward bias) with sufficient amplitude to reduce the collector I a little, causes a sudden change of I through the inductance, which induces a voltage across this inductance. This voltage adds to the collector supply potential. The effect is to increase the *instantaneous* collector potential to a value considerably above the operating potential and hence above the collector-supply potential.

With a swing in the opposite direction for the input signal, forward bias is increased, and a sudden I change occurs again in the transformer primary inductance. Now, a voltage is induced across the primary inductance that is opposite to the previous change. Thus, the collector-voltage signal swing reaches values above and below the normal operating potential.

As long as the average dc collector current does not change when the signal is applied to the base input circuit, the same dynamic characteristics hold for the resistance loads as for the transformer-coupled loads.

Note that the constant power-dissipation line is indicated as 13.2 W and that the load line is close to it for obtaining maximum power output. Using either Eq. (3-1) or Eq. (3-2), we find that the load line has a value of 15.9 Ω and hence is sufficiently low to permit direct coupling to a voice coil of 16 Ω.

$$R_L = \frac{dV_c}{dI_c} = \frac{(0 \text{ to } 28)}{(0 \text{ to } 1.76)} = 15.9 \, \Omega$$

The 1.76 is an approximate value but is sufficiently close for practical purposes. Thus, the load resistance can also be considered to be 16 Ω.

The actual signal-power output depends on the magnitude of the input signal and how nearly it causes a full swing of collector current. The load is resistive; if not an actual resistor, it is the resistance reflected back to the transistor by the output transformer. Thus, signal-power output is found by multiplying the signal E by I. To this end, consider that the maximum value of the sine wave signal output voltage is

$$E = \frac{E_{max} - E_{min}}{2} \tag{3-3}$$

However, the rms value of a sine wave signal is found by dividing the voltage maximum by the square root of 2:

$$E_{rms} = \frac{E_{max} - E_{min}}{2\sqrt{2}} \tag{3-4}$$

Similarly, for sine wave signal current, the rms value is

$$I_{rms} = \frac{I_{max} - I_{min}}{2\sqrt{2}} \tag{3-5}$$

Hence, multiplying Eq. (3-4) by Eq. (3-5) produces the equation for finding signal-power output:

$$P_{out} = \frac{(E_{max} - E_{min})(I_{max} - I_{min})}{8} \tag{3-6}$$

If, for instance, the collector voltage changes from 4 V to 20 V during the presence of an input signal with a consequent current change (along the load line of Fig. 3-4) of from 0.5 to 1.5 A, Eq. (3-6) produces the following:

$$\frac{(20 - 4)(1.5 - 0.5)}{8} = \frac{16}{8} = 2 \text{ W}$$

The change in power output for an increase in input-signal drive can be compared if we assume that sufficient input-signal drive was applied to cause the collector voltage to change from 0 to 24 V, with an accompanying current change of 0.25 to 1.76 A:

$$\frac{(24 - 0)(1.76 - 0.25)}{8} = \frac{36.24}{8} = 4.53 \text{ W}$$

The percentage of harmonic distortion should be calculated in relation to the maximum signal swing, because low values of input signals mean operation on a more linear portion of the characteristic curves and are not a true indication of what occurs at higher input-signal levels. Similarly, the distortion should be calculated on the basis of a pure sine wave input signal, where a base-current change of equal amplitude occurs on each side of the base-current operating point.

Thus, for Fig. 3-4, with an equal change of I_b above and below the operating point, we can obtain values of collector-current changes. For the operating point at 30 mA, assume a change to 10 mA in one direction, and from 30 mA to 50 mA for the signal swing in the opposite direction. The corresponding I_c changes are then used in the following equations. Either of these solves for second-harmonic distortion, which is the primary type in triode transistors. Hence, the percentage of second-harmonic distortion can be used as an approximation for total harmonic distortion.

Percentage of second-harmonic distortion

$$= \frac{(I_{max} + I_{min}) - 2I_o}{2(I_{max} - I_{min})} \times 100 \tag{3-7}$$

Power Amplifier Load Line Factors / 55

In the foregoing the I_o indicates the collector current operating point; for Fig. 3-4 this occurs at the -1 A value. Instead of Eq. (3-7), the following can also be used:

Percentage of second-harmonic distortion

$$= \frac{[\frac{1}{4}(I_{max} + I_{min})] - \frac{1}{2}I_o}{\frac{1}{2}(I_{max} - I_{min})} \times 100 \qquad (3\text{-}8)$$

For Fig. 3-4, at 10 mA for I_b the collector current minimum is 0.3 A, whereas at 50 mA for I_b the I_c maximum is 1.62 A (approximately). Using the collector current values thus indicated in Eq. (3-7), we obtain

$$\frac{(1.62 + 0.3) - (2 \times 1)}{2(1.62 - 0.3)} = \frac{1.92 - 2}{2 \times 1.32} = \frac{-0.08}{2.64} \times 100 = 3\%$$

For Eq. (3-8) we obtain the same percentage of distortion:

$$\frac{[\frac{1}{4}(1.62 + 0.3)] - 0.5}{\frac{1}{2}(1.62 - 0.3)} = \frac{-0.02}{0.66} \times 100 = 3\%$$

The percentage of collector efficiency can be found by the following equation, based on output-input power ratios:

$$P_{eff} = \frac{\text{signal-power output}}{\text{dc supply power}} \times 100 \qquad (3\text{-}9)$$

Because variations in signal amplitudes would alter loading effects on the amplifier, Eq. (3-9) should be applied only to constant-amplitude sine-wave-type test signals that apply a constant load on the amplifier. For Fig. 3-4 the power supply furnishes 12 W of power at I_o because we have 1A and 12 V. Relating this value to the 4.5 W obtained earlier, we find that the percentage of collector efficiency is

$$\frac{4.5}{(1 \times 12)} = \frac{4.5}{12} \times 100 = 37.5\%$$

Once the required power output has been calculated and the design has been slanted toward the minimum distortion factors with best collector efficiency, the circuit can be set up and calculations verified by using commercial distortion analyzers, power meters, and similar equipment. The slant of the load line can then be altered slightly to compensate for any variations or deviations from the performance values that are planned.

While transformer ratios and coupling factors as subsequently discussed can be mathematically analyzed, transformer characteristics may alter initial calculations concerning power output, efficiency, and distortion; hence the circuit may have to be set up and results checked with the particular transformer to be used. Additional details are covered in the remaining sections of this chapter.

3-4 TRANSFORMER FACTORS

Signal attenuation in transformers has already been mentioned in Sec. 3-1. Loss characteristics are reasons for the reluctance of design engineers to use transformer coupling between stages unless it is necessary to do so in terms of cost factors or objectionable alternates. In high-fidelity amplifiers the preferred method is to use direct coupling between stages. For inexpensive amplifiers in portable receivers, however, it is often more expedient to use a transformer to realize simplicity in impedance matching (and voltage gain in step-up interstage units). Also, transformers are used in some signal-generator devices, in certain television circuitry, and also in industrial control electronics. Hence, loss factors and basic functional aspects must be considered in design practices.

A transformer is assumed to have virtually no internal resistance. If the resistance of the primary winding is appreciable, some signal power will be consumed in the primary and hence wasted. This is undesirable, because as much as possible of the power generated within the transistor should develop across the load, such as the *voice-coil impedance*. To obtain a minimum of dc resistance, the primary winding can be made of fairly large wire. Actually, the wire must have a diameter sufficiently large to carry the collector current without undue overheating. An overheating of the transformer indicates that the internal resistance is consuming considerable power, and such power, whether audio signal energy or dc power from the power supply, is a waste of energy. When the primary has such a low resistance that it is a negligible factor, the *signal power* circulating in the primary is transferred into the secondary with the greatest efficiency. A well-designed transformer has an efficiency of well over 95 percent.

The primary of the output transformer *is not* the actual load resistance; neither is the secondary. Rather, the audio signal power is applied to the voice coil of the speaker, and the resultant signal current sets up varying magnetic fields. The latter aid or oppose the fields of the speaker magnet, causing the loudspeaker diaphragm cone to vibrate and produce audible sounds. Thus, the electric energy of the power amplifier is converted into acoustical energy. It must be noted that the output transformer must match the load (such as the voice coil) to the *load line resistance of the transistor.*

If the transformer has an appreciable distributed capacitance, some of the high-frequency audio signal components will be attenuated. For this reason, the transformer design should be such that a minimum of distributive capacitances are present. For the same reason, the wire size should be held at a minimum, because a larger wire size contributes to capacitance effects between adjacent turns, as well as between adjacent layers of the transformer. (See Sec. 3-1.)

Other factors in transformer efficiency are the quality and quantity of the laminated core. A larger core increases permeability, and hence the number of turns in the primary and secondary can be reduced, while the required amount of inductance is still held constant. The reduced number of turns means less distributed capacitance, as well as a decrease in the internal resistance of the transformer. Consequently, less signal energy is shunted by distributed capacitances, and less signal energy is wasted because of the series dc resistance of the primary winding. The dc resistance consumes not only audio signal-energy power (which is ac), but also dc power from the power supply.

Another factor to be considered is the variation of inductive reactance with changes of signal frequency, as mentioned earlier. A low audio signal frequency decreases the reactance of the primary, and hence less signal voltage develops across it than is the case for higher signal frequencies. This tends to attenuate the lower-frequency audio signal components, and upsets uniform response. Good transformer design, however, helps keep such variations to a minimum, as well as reducing losses for the higher audio signals by decreasing the shunting effect caused by distributed capacitances within the transformer.

Although such design considerations increase signal frequency response and transformer efficiency, their adoption is not always compatible with cost factors, because each improvement in transformer design increases the manufacturing costs. Thus, the design engineer must often compromise with regard to the final selection to obtain the best performance within the framework of the costs involved so as to maintain competitive advantages in the final marketing of the systems containing the transformer.

3-5 COUPLING RATIOS

The degree of coupling between the primary and secondary of a transformer determines the efficiency of signal-power transfer. The *co-*

efficient of coupling is expressed as k, and unity is only achieved when the secondary winding is fully capable of intercepting virtually all of the magnetic lines of force of the primary winding. This is the case when the secondary winding is wound around or over or is interwound with the primary winding. Under such conditions practically all of the lines of force cut both the primary and secondary windings, and this is usually the case with the nonresonant low-frequency transformers. With the resonant types covered in Chapter 4, coupling varies considerably, as discussed in Secs. 4-17 and 4-18.

Because the coefficient of coupling is the ratio of the flux linking the two windings to the flux established by the primary winding current, *unity coupling* means that the amplitude of the electric energy available from the secondary winding of the transformer is proportional to the ratio of turns between primary and secondary. Voltage is directly proportional to the number of turns of the secondary with respect to the primary. Hence, if the secondary winding has fewer turns than the primary, the voltage will be *stepped down*. If the number of turns in the secondary is greater than in the primary, the voltage will be *stepped up* in proportion to the increase in the number of turns in the secondary. In equation form, this is expressed as

$$\frac{E_p}{E_s} = \frac{N_p}{N_s} = a \qquad (3\text{-}10)$$

where E_p and E_s are the primary and secondary voltages.

N_p and N_s are the number of primary and secondary turns.

E_p/E_s is the voltage ratio.

N_p/N_s is the number of turns ratio.

a is the *transformation ratio*.

The number of turns in the primary of a transformer is determined by the amount of *no-load current* that is to flow. As more turns are added to the primary, the impedance rises and the opposition to the flow of ac increases. In a well-designed transformer only a negligible amount of power is used with no load connected to the secondary. As an increasing load is impressed on the secondary, more current flows, and greater power is drawn from the source.

Example: The secondary of a transformer is to deliver 10 V. The primary has 720 turns and the ac source is 120 V. How many turns must the secondary winding have?

Answer:

$$a = \frac{E_p}{E_s} = \frac{120}{10} = 12$$

$$N_s = \frac{N_p}{a} = \frac{720}{12} = 60 \text{ turns}$$

Example: A high-voltage transformer for producing an arc in an oil burner is to operate on 120 V and deliver 15,000 V at the secondary. What is the turns ratio? If the primary has 600 turns, how many turns must the secondary winding have?

Answer:

$$\frac{N_p}{N_s} = \frac{120}{15,000} = 0.008 \text{ to } 1 = 600 \text{ to } 75,000$$

Because this is a ratio, the secondary versus the primary turns could have been calculated instead, producing the same turns ratio:

$$\frac{N_s}{N_p} = \frac{15,000}{120} = 125 \text{ to } 1 = 75,000 \text{ to } 600$$

The voltage in relation to a single turn of either the primary or secondary winding is known as the *volts per turn* (vpt), and for either the primary or secondary is indicated by the equation

$$E_{pt} = \frac{E}{N_t} \qquad (3\text{-}11)$$

If both the vpt and the number of turns in the secondary are known, the voltage at the secondary is the product of the vpt and the secondary turns:

$$E_s = E_{pt} \times N_s \qquad (3\text{-}12)$$

Example: A transformer has 1000 turns in the primary and 5.25 turns in the secondary. The primary voltage is 120. What is E_{pt} and E_s?

Answer:

$$E_{pt} = \frac{120}{1000} = 0.12 \text{ vpt}$$

$$E_s = 0.12 \times 5.25 = 6.3 \text{ V}$$

3-6 Z MATCHING

When the generator or other source of ac signal does not match the load impedance, a transformer is an ideal method for achieving an impedance match, because the transformer can be utilized to *step up or step down impedances* as well as voltages. When it thus produces an impedance match, a maximum of power is delivered to the load. The maximum value depends, of course, on the amount of voltage applied to the load and the amount of load current drawn.

The signal power available from the secondary windings is equal to the power that may be drawn from the ac generator or power mains furnishing energy to the transformer. Thus, if the transformer steps up the voltage, the available current will be less. Also, if a step-down transformer is used, there will be less voltage at the secondary, but greater available current (thus again furnishing power approximately equal to that available from the source). The transformer itself must have a wire size and number of turns that permit it to handle the amount of power required.

It is evident that the ratio of voltages equals the ratio of currents:

$$\frac{E_p}{E_s} = \frac{I_s}{I_p} \tag{3-13}$$

Thus, from Eq. (3-10) in relation to the foregoing, we have these relationships:

$$\frac{I_s}{I_p} = \frac{N_p}{N_s} = a \tag{3-14}$$

and

$$Z_p = \frac{E_p}{I_p} \tag{3-15}$$

The power consumed by a load transducer attached to the secondary winding is reflected into the primary because of the additional current flowing through the primary. The impedance of the load is, in effect, also reflected back to the primary and appears as though it were shunting the primary. Its impedance can be found by Eq. (3-15). From Eq. (3-12) we can ascertain the impedance factors in relation to the primary and secondary windings.

$$E_p = \frac{N_p}{N_s} E_s = aE_s \tag{3-16}$$

and

$$I_p = \frac{N_s}{N_p} I_s = \frac{I_s}{a} \tag{3-17}$$

Relating the two, we have

$$Z_p = Z_s \left(\frac{N_p^2}{N_s}\right) \tag{3-18}$$

Thus, when a load is connected across the secondary, its impedance is reflected back across the transformer to the primary, and hence Z_p is often referred to as the *reflected impedance*. With a transformer having unity coupling, the impedance ratio is equal to the square of the turns ratio, as indicated by Eq. (3-18). Since the expression $(N_p/N_s)^2$ is the transformation ratio squared, and since the secondary impedance Z_s consists mainly of the load impedance, Eq. (3-18) can be expressed as

$$Z_p = a^2 Z_L \tag{3-19}$$

$$\therefore \ a^2 = \frac{Z_p}{Z_L} \tag{3-20}$$

The turns ratio necessary to match two impedances can then be expressed as

$$a \text{ (turns ratio)} = \sqrt{\frac{Z_p}{Z_L}} \tag{3-21}$$

Example: A transistor audio amplifier requires a 2400-Ω load impedance. The speaker has an impedance of 8 Ω. The primary has 865 turns. How many turns must the secondary winding contain?

Answer:

$$a = \sqrt{\frac{2400}{8}} = \sqrt{300} = 17.3 \text{ to } 1$$

$$N_s = \frac{N_p}{a} = \frac{865}{17.3} = 50 \text{ turns}$$

Example: The turns ratio of a transformer was tested by applying 10 V to the secondary and reading the primary voltage to find the voltage ratio. The primary voltage reads 200 V. If a 400-V generator is to be connected to the primary, what is the load impedance for a match?

Answer:

$$\frac{200}{10} = 20 \text{ to } 1 \text{ turns ratio}$$

Turns ratio squared $(a^2) = 20 \times 20 = 400$

$$a^2 = \frac{Z_p}{Z_L} = \frac{4000}{Z_L}$$

$$\therefore 400 = \frac{4000}{Z_L}$$

$$Z_L = \frac{4000}{400} = 10 \, \Omega$$

3-7 PUSH-PULL MATCHING

When two transistors in a push-pull circuit connect to an output transformer as shown in Fig. 3-5, the resistance represented by the actual load (the voice coil of a speaker in this instance) must again match the representative load resistance established by the load line used for the transistors. (As mentioned earlier for single-ended stages, direct coupling is preferred to minimize the losses contributed by the transformer. The output-transformer system is, however, encountered in many devices where exceptional high fidelity is not a requisite.)

For the circuit shown in Fig. 3-5, the reflected load from the speaker resistance is present at the collector of each transistor. If we designate R_L as the indicated load resistance established by the load line

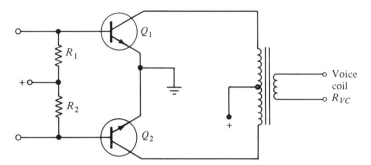

Figure 3-5 Push-pull output transformer

for each transistor, N_p as the number of turns in the total primary winding, n_s as the secondary turns, and R_{vc} as the voice-coil resistance, R_L may be solved by

$$R_L = \frac{1}{4}\left(\frac{N_p}{N_s}\right)^2 R_{vc} \qquad (3\text{-}22)$$

Thus, for a designated load line R_L and a selected value of voice-coil load resistance R_{vc}, we can solve for the necessary turns ratio of the output push-pull transformer by

$$a_{pp} = 2\sqrt{\frac{R_L}{R_{vc}}} \qquad (3\text{-}23)$$

Thus, if each transistor in a push-pull circuit has a load line indication of 1600 Ω, and the voice coil = 8 Ω, the turns ratio is

$$2\sqrt{\tfrac{1600}{8}} = 2\sqrt{200} = 2 \times 14.2 = 28.4 \text{ to } 1$$

Thus the turns ratio of 28.4 to 1 could mean 568 turns in the primary versus 20 in the secondary, or 852 in the primary versus 30 in the secondary, etc. (The number of turns indicated for the primary refers to the *total* primary from the collector of Q_1 to that of Q_2.)

Power output for push-pull is approximately twice that obtained for a single transistor for a given set of operating conditions. If there is a reduction in the generation of undesired harmonics, there will be more than double the undistorted output power, because the harmonic components rob the total output power of some of its undistorted signal amplitude.

In push-pull operation the signal swing of one transistor with a given polarity adds to that of the other transistor of opposite polarity. Because of the phase difference at the base input of each transistor, current increases through one collector while decreasing in the other. The span of the positive change to the negative constitutes the total *signal change* that induces energy to the secondary winding and provides double the signal power output from the push-pull circuit. Thus, if Eq. (3-6) for power output is to be used, the double span of the push-pull must be taken into consideration. For the example previously given, the E_{max} of 24 V to 0 would now be 48 V, and the I span (1.76 to 0.25, a total of 1.51) would now be 3.02. When the products are divided by 16 (instead of 8), we obtain double the output that was the case earlier:

$$P_{out} = \frac{(48)(3.02)}{16} = 9 \text{ W}$$

3-8 DECOUPLING CIRCUIT

In all amplifier circuits, the common power supply (or battery) serves as an interconnection between signals of the various stages, and in consequence adverse feedback of signals can occur. To minimize the interaction that results from common power sources, a decoupling network is utilized to isolate signal voltages between amplifier stages. A properly designed decoupler circuit also extends the signal-frequency response of the amplifier by providing low-frequency compensation. Also, the series resistor of the network can aid in establishing the proper dc potentials to the output circuit of the transistor amplifier. Decoupling networks are found extensively in various audio and RF amplifier systems.

Figure 3-6(a) shows an output amplifier coupled to a loudspeaker by an interstage transformer. Capacitor C_1 bypasses very high frequency signals to minimize high-frequency noise. The decoupling network consists of capacitor C_2 and resistor R_2. The capacitor has a low reactance shunting effect for signals, particularly those of higher frequency, and hence minimizes common coupling in the power supply. Because of its low reactance, this capacitor also places the bottom of the transformer primary at signal ground.

In Fig. 3-6(b) a similar decoupling network is utilized with the transistor's load resistor. Here, the decoupling network consists of resistor R_3 and capacitor C_3. Resistor R_2 is the collector load resistor across which the amplified signals develop for coupling to the subsequent

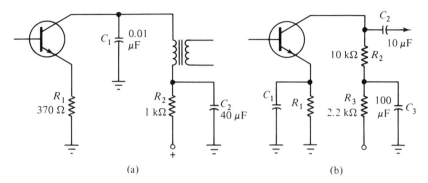

Figure 3-6 Decoupling networks

stage. As with the network in (a), capacitor C_3 not only shunts R_3 but also places the bottom of R_2 at signal ground because of its low reactance. Since reactance values depend on frequency, C_3 has a lower capacitive reactance for signal frequencies that are higher. Consequently, the decoupling network acts as a base-compensating system, because it increases or decreases the effective load resistance across which signals develop. For higher frequencies, for instance, where C_3 has a very low reactance, resistor R_3 is substantially shunted, and virtually all the signal voltages appear across the decoupling resistor R_2. Similarly, the coupling capacitor C_2, which couples the signal energy to the next stage, also has a very low reactance for signals of higher frequencies. Thus, most of the signal energy is coupled to the subsequent stage.

For low-frequency signals, the reactance of C_2 increases, and the amplitude of the signals applied to the next stage declines. Thus, the lower frequency signals are attenuated. The decoupling network compensates for this to some extent because, for lower signal frequencies, the reactance of C_3 rises, and the decoupling capacitor consequently has less shunting effect for resistor R_3. When this occurs, some of the signal components develop across resistor R_3 as well as the load resistor R_2. This has the effect of placing the two resistors in series, increasing the total load resistance value and thus raising the amplitude of the voltage drop across the load resistor for the signals. Consequently, the signal amplitude is increased to compensate for the attenuation caused by coupling capacitor C_2. Thus, the decoupling network has a variable characteristic that changes as the frequency of the signals change.

In practical circuitry the ohmic value of the decoupling resistor R_3 is selected so that its value is approximately one-fifth that of R_2 and about ten times higher in ohmic value than the reactance of C_3 for the lowest frequency encountered in the amplifier stage. For the audio amplifier coupling shown in Fig. 3-6(b), the values given are typical. For RF amplifiers, capacitor values would be much smaller, since lower reactance values prevail for the signals of higher frequencies.

3-9 INVERSE FEEDBACK

Inverse feedback is used between several stages of an amplifier to reduce harmonic distortion, increase the bandpass, reduce noise signals, and improve stability. With inverse (negative) feedback, signal amplification is reduced in proportion to the amount of signal fed back, but this disadvantage is outweighed by the improved performance.

Inverse signal feedback utilizes a return circuit from an output stage to an input stage, but is so designed that the signals that are fed back are 180 deg out of phase with the signals prevailing at the feedback stage. (Additional data on feedback, plus appropriate mathematics, will be found in Sec. 6-14, "Operational Amplifier.")

A representative inverse feedback loop is shown in Fig. 3-7. Here, the signal that is fed back is applied across the emitter resistor of the initial amplifier stage. The signal is obtained from the output amplifiers as shown. The amount of feedback is regulated by the selection of the appropriate value resistor-capacitor components in the feedback line. The values given in Fig. 3-7 are for reference purposes, but the specific values depend on the degree of feedback that is required, plus the amount of amplification reduction which can be tolerated.

The feedback signal that appears across the 500-ohm emitter resistor of the input stage acts inversely to the input signal of that stage.

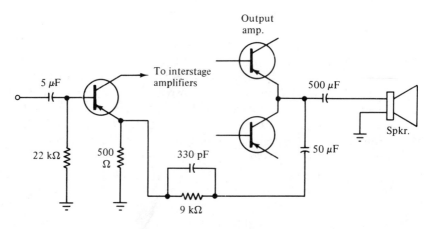

Figure 3-7 Inverse feedback

Thus, for a positive alternation of the input signal, a certain amplitude negative alternation occurs in the collector circuit, but the inverse feedback signal will change the forward bias between base and emitter and reduce the amplitude of the negative alternation of current circulating in the emitter-collector circuit. Similarly, for a negative alternation of the input signal, the positive alternation of the amplified signal would develop in the emitter-collector circuit, but the amplitude would be reduced below that which would prevail without a feedback. (It must be remembered that there is a phase difference between the signal applied to the base of the transistor circuit and the amplified version that develops at the collector side in a common-emitter circuit.)

The 50-μF capacitor in the feedback line blocks the dc present at the collector-emitter of the output amplifier from appearing across the 500-ohm resistor of the input amplifier. The 9-kΩ resistor and its shunting capacitor (330 pF) determines the amount of signal fed back.

When the emitter resistor does not have a bypass capacitor across it, as is the case for the circuit shown in Fig. 3-7, another form of inverse feedback occurs. Such feedback, sometimes termed *current* type, also has a degenerative effect on the amplification of the particular stage involved, but the unbypassed circuit provides for a better frequency response, as discussed more fully in Sec. 1-8.

3-10 DIRECT COUPLING

As illustrated earlier in Fig. 3-1, a coupling capacitor has a varying reactance for signals having different frequencies, and consequently the response of the amplifier to lower-frequency signals is diminished. Faults contributed by coupling capacitors can be eliminated by using *direct coupling* of the type shown in Fig. 3-8. Here, the collector output

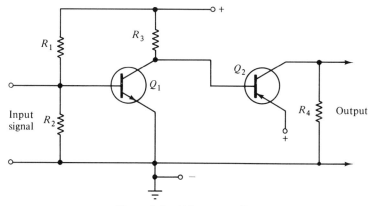

Figure 3-8 Direct coupling

from transistor Q_1 connects directly to the base input of the transistor Q_2.

When transistors are coupled directly in this manner, the problem arises with respect to the allocation of proper bias polarities if the transistors are identical. By the use of complementary transistors, however, voltage distributions are simplified. For the circuit shown in Fig. 3-8, an *npn* transistor is used, and thus the emitter should be made negative with respect to the base to provide the necessary forward bias at the input. This requirement is accomplished by the voltage divider resistors R_1 and R_2. Resistor R_3 is the load resistor for the collector of Q_1 and for the necessary reverse bias; the supply potential is positive as shown.

For Q_2, a *pnp* is used, and the positive emitter potential is applied as shown. For the necessary forward bias at the input, it would appear that the base of Q_2 is positive, since it connects to the positive supply terminal because of the direct coupling to the collector of Q_1. Resistor R_3, however, in conjunction with the impedance of transistor Q_1 (emitter collector) also shunts the supply potential, and thus forms a voltage divider with the Q_1 impedance. Hence, the junction of R_3 and the collector of Q_1 is negative with respect to the positive terminal of the battery or power source. Thus, the base of Q_2 is negative with respect to the positive emitter by a specific voltage value. Hence the input bias requirements for both transistors are satisfied with this complementary circuitry. Resistor R_4 is the load resistor across which the output signal voltage develops, and this resistor also couples the necessary negative potential for the collector of Q_2 by being connected to the negative ground line as shown.

3-11 DESIGN-PROCEDURE VARIABLES AND CONSIDERATIONS

Junction-transistor design data can be obtained from characteristic curves, as illustrated earlier in this chapter (Figs. 3-2, 3-3, and 3-4) for calculating such factors as gain for a given load line impedance and signal swing, comparison between operating point characteristics, and the performance in relation to the slope of the load line. Such information is often furnished by the manufacturer for a given transistor, so curves would not have to be plotted. Thus, from data supplied on a particular transistor's characteristics, circuit parameters can be ascertained and component values calculated.

The component values that are finally selected depend on the type of circuit to be designed, the operating voltages selected within the limits specified by the transistor information sheet, and the compromises that must be considered to balance the cost factors with the best performance that can be obtained while imposed restrictions are observed. In building equipment for our personal use, we may elect to use 100-μF filter capacitors instead of 50 μF and may prefer a working voltage rating of 35 V when the actual working voltage is much less. The higher ratings, however, give us added protection against burnout and provide better filtering. Thus, if cost (and physical sizes) is no factor, better systems can be designed and superior components can be selected, including transistors with better-than-average ratings and characteristics.

For commercial gear we must initially ascertain the market level for which the device is aimed. If, for instance, a medium- or low-priced stereo amplifier is under consideration, we may be satisfied with an rms audio-power rating per channel of 10 or 20 W and may even tolerate harmonic distortion of close to one percent. For better quality gear we may select transistors and power supplies to furnish 40 or 60 W (or more) per channel, with a distortion factor of less than 0.2 percent. Hence, all these factors are relevant in the design of any electronic gear, and no rigid rules can be given that a particular load resistor must be 6 kΩ for a specific amplifier, or that a bypass capacitor must have a shunt reactance less than 200 Ω for the lowest frequency signals to be handled.

Besides such variables, somewhat different though fairly close values for components may be obtained by the selection of the equation or procedures utilized to obtain a given result. Consequently, the more data regarding procedures, methods, equations, and related information the designer has on hand, the more he will be able to expedite the solution of a particular design problem or to undertake a major design program for a complete electronic system with greater facility. Consequently, in addition to the discussions given in earlier sections of this chapter (and in earlier chapters), a number of related equations, explanations, and examples are furnished in the remaining sections of this chapter for the transistor, the JFET, and the MOSFET.

3-12 DEFINITION OF LETTER SYMBOLS

The following comprises a list of transistor letter symbols in general usage. These summarize some of those that were defined earlier in Sec. 2-7. Several others have been added for convenience when reference to a more complete list is required.

h_{fe} Common emitter
h_{fb} Common base   } Small-signal forward-current transfer ratio
h_{fc} Common collector

h_{FE} Common emitter dc forward-current transfer ratio
h_{FB} Common base dc forward-current transfer ratio
h_{ib} Common base ⎫
h_{ic} Common collector ⎬ Small-signal input impedance
h_{ie} Common emitter ⎭
h_{oe} Common emitter ⎫
h_{ob} Common base ⎬ Output admittance
h_{oc} Common collector ⎭
h_{re} Common emitter ⎫
h_{rc} Common collector ⎬ Small-signal reverse voltage transfer ratio
h_{rb} Common base ⎭
I_c Collector current (rms)
i_c Collector current (instantaneous)
I_b Base current (rms)
i_b Base current (instantaneous)
I_e Emitter current (rms)
i_e Emitter current (instantaneous)
BV Breakdown voltage
V_{EB} Emitter-base voltage
V_{CE} Collector-emitter voltage
V_{CB} Collector-base voltage

In the foregoing list the subscript letters that are capitalized refer to static or dc conditions, such as the h_{FE} and the h_{FB}. (On occasion the first letter has been in upper case also, as H_{FE}. The lowercase subscript letters refer to ac-type signal-current ratios (h_{fe}, h_{fb}, etc.). The alpha [Eq. (2-4)] and beta [Eq. (2-1)] given earlier refer to the small signal-current ratios as well as the static dc ratios. Thus, alpha is h_{fb} or h_{FB} and beta is h_{fe} or h_{FE}.

3-13 DESIGN-RELATED EQUATIONS AND APPLICATIONS
(Part 1—Junction Transistors)

The most simple bias system for the transistor is shown in Fig. 3-9, where a series resistor R_B is used to provide the necessary voltage drop from V_{CC} (the supply potential) to the base terminal. The bias circuit can be considered as R_B in series with the input resistance of the transistor, since electron flow is from the negative terminal V_{CC} through R_B and through the transistor to ground. Hence, the value of R_B depends on the amount of bias voltage required, the amount of current flow in the

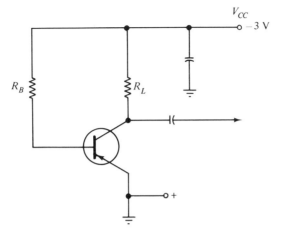

Figure 3-9 Basic bias system

base circuit (at the operating point selected), and the value of the supply voltage V_{CC}. Hence, the ohmic value of R_B can be found by the following equation:

$$R_B = \frac{V_{CC} - \text{bias voltage}}{I_b} \qquad (3\text{-}24)$$

Thus, if the base current $I_b = 0.0001$ A and the bias needed is 0.2 V, the value of $R_B = 28{,}000 \ \Omega$:

$$\frac{3 - 0.2}{0.0001} = 28{,}000 \ \Omega$$

The results here are somewhat inflexible, which is a disadvantage if we need to match the impedance of the input to the transistor with that of the circuit feeding in the signal. Also, greater circuit stability can be achieved by using a voltage-divider system for bias, as shown in Fig. 3-10, consisting of R_1 and R_2, which bridges the supply source V_{CC}.

For Fig. 3-10 additional stability is achieved by using an additional resistor (R_3) in the emitter circuit. For any temperature change that may cause an increase in current through the transistor, there would be a corresponding rise in voltage across R_3. Thus, the voltage drop across R_3 reduces the forward bias at the base, since the emitter becomes more positive than the base. The bias change reduces current through the emitter-collector circuitry and hence compensates for the change that developed because of thermal effects. Total bias is the difference between the E across R_2 and R_3.

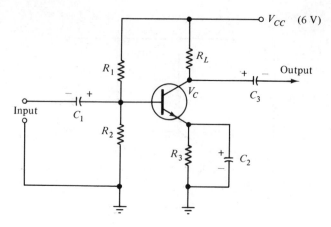

Figure 3-10 Bias with improved stability

For the circuitry shown in Fig. 3-10 various values for R_1 and R_2 can be selected as needed for impedance matching, circuit stability, and optimum current-transfer ratios. Resistor R_2 should not be too low a value, because less input signal voltage will develop across it. Also, however, if R_2 has too high a value, less current flows through it in proportion to the shunt current flowing through the transistor base circuit. Hence, stability is affected, since current flow through R_1 is primarily I_B.

In the design of the circuit shown in Fig. 3-10 other factors must be considered in the selection of component values. When input and output impedances are given, the ohmic values of R_2 and R_L should be as close to these as possible for maximum signal transfer into and out of the circuit. Some compromises may, however, be necessary to achieve desired balances between circuit stability and signal gain. If the circuit gain (at a minimum) is to be 10 with average stability, R_2 should be 10 times higher in ohmic value than the emitter resistor R_3. This ratio is termed the *stability factor* and is equal to the value of R_2 divided by R_3. If R_2 is made higher than R_3 by a factor of 20, gain will be high but stability low. A factor of 5 would produce very high circuit stability, but low signal gain. Thus a compromise must be reached but without going higher than 20 or below 5 as acceptable limits.

For class A operation of the circuit in Fig. 3-10 signals must operate on the linear portion of the characteristic curves and should not swing too near to the limits of saturation and cutoff. Preferably the collector should be operated at a voltage (V_C) one-half that of the source voltage V_{CC} for maximum positive and negative signal swing at the collector, and minimum signal distortion.

The value of R_L must also be selected as a compromise between maximum gain and signal transfer versus stability. Preferably, the ohmic value of R_L should always be less than 20 times the value of R_3, but at least five times the ohmic value of resistor R_3. Since the values of R_3 may range from 100 Ω to 1500 Ω, the R_L ohmic value may lie between 500 Ω and 7500 Ω.

As an example of the calculations involved for Fig. 3-10, assume that the base current at the operating point is to be 0.0001 A and R_2 is to have a value of 2500 Ω to match the input impedance. Bias applied to the base terminal from across R_2 is to be 0.2 V. Under these conditions, what must the value of R_1 be?

The base current flow plus the current flow through R_2 also flow through R_1. Thus, the total current through R_1 branches so that 0.0001 A flows through the input circuit of the transistor. The current through $R_2 = E_2/R_2$, or 0.2/2500, which is 0.00008 A, or 0.08 mA. Because this current and the base current flow through R_1, the voltage drop across $R_1 = IR$ and hence is

$$E_{R_1} = (I_2 + I_{em})R_1 \quad \text{and} \quad E_{R_1} = V_{CC} - V_{BE} \qquad (3\text{-}25)$$

Hence,

$$R_1 = \frac{V_{CC} - V_{BE}}{(I_2 + I_{em})} \qquad (3\text{-}26)$$

Inserting voltage and current values gives us the following:

$$R_1 = \frac{6 - 0.2}{(0.00008 + 0.0001)} = \frac{5.8}{0.00018} = 32{,}222 \; \Omega$$

The following proves that 5.8 V drops across R_1 and the required 0.2 V at the base input:

$$E_{R_1} = IR_1 = 0.00018 \times 32{,}222 = 5.799 \text{ V or } 5.8 \text{ V}$$

V_{CC} of $6 - 5.8 = 0.2$ V at base terminal

If R_2 had been specified to have a value of 1000 Ω the following would have given us the new value for R_1:

$$I_{R_2} = \frac{0.2}{1000} + 0.0002$$

$$\text{and} \quad R = \frac{5.8}{0.0001 + 0.0002} = \frac{5.8}{0.0003} = 19{,}333 \; \Omega$$

For the original 2500-Ω value assigned to R_2, resistor R_3 in the emitter should be about one-tenth the value of R_2 for a good stability factor. Hence, R_3 consists of a 250-Ω resistor. For R_L a value was chosen that is five times that of R_3; hence R_L has a value of 1250 Ω.

Capacitor C_1 and resistor R_2 are in series (to ground), and hence the input signal voltage drops across the X_C of C_1 and the resistance of the base resistor R_2. Hence, the base terminal of the transistor has a signal voltage applied that is obtained only from that which drops across the R_2 portion of the total signal amplitude input. Thus, the X_C should be sufficiently low in ohmic value to cause minor or negligible attenuation of signals at the low-frequency end of the frequency response of the audio (or video) band. To provide for a low reactance, the low-frequency cutoff point for C_1 must be taken into account. The cutoff is that signal frequency that causes the X_C of C_1 to have a value one-half of the ohmic value of R_2. Because in our example R_2 is 2500 Ω, the reactance of C_1 should be at least as low as one-half of R_2 or lower. Too large a capacitor, however, becomes more costly, and some stray capacitance may be established between the larger bulk of the coupling capacitor and adjacent components.

Selecting 40 Hz as the low-frequency reference and 1250 Ω as one-half of R_2 for Fig. 3-10, we obtain

$$C = \frac{1}{(6.28 f X_C)} = \frac{1}{6.28 \times 40 \times 1250} = 3 \,\mu\text{F}$$

Since a 5-μF capacitor with the same voltage rating as the 3-μF capacitor does not contribute materially to bulk or cost, it can be used to lower the reactance additional at 40 Hz:

$$X_C = \frac{1}{(6.28 f C)} = 796 \,\Omega$$

Thus, only a fractional signal-voltage amplitude develops across the reactance of C_1 of 796 Ω as against the 2500 Ω of resistance for R_2. Thus for 40 Hz only about one-third of the signal develops across C_1 and two-thirds appears across R_2, while for successively higher signal frequencies the drop across C_1 halves for every doubling of the 40-Hz frequency; hence virtually all signal amplitudes appear across R_2 for higher frequencies.

Capacitor C_2 shunts R_3 and should have a reactance substantially lower than the resistance of R_3 to provide for effective bypassing of signal components around R_3 and for minimizing degeneration. Since $R_3 = 250 \,\Omega$, one-fifth this value for X_L would provide an effective signal shunt around R_3. The capacitor value, therefore, is

$$C = \frac{1}{6.28 \times 40 \times 50} = 80\,\mu\text{F}$$

A 100-μF capacitor can be used to increase the shunt effect slightly (40 Ω).

With resistor R_3 bypassed, some variations may occur in the value of the input impedance, though gain is increased over the unbypassed method. Without the bypass, however, a form of inverse feedback occurs (current type), which aids in distortion reduction and also improves frequency response. Generally, if the gain increase can be tolerated, it is advisable to leave the resistor unbypassed.

For the output coupling capacitor C_3, similar factors prevail as was the case for C_1. Capacitor C_3 is related to the load resistance R_L, and if the latter is 1250 Ω, half this value for X_C at the low-frequency cutoff would be satisfactory. Thus, a 625-Ω X_C requires a capacitor having a value as determined by the following:

$$C = \frac{1}{6.28 \times 40 \times 625} = 6.37\,\mu\text{F}$$

Thus, a 10-μF capacitor standard value can be used to provide for even a lower capacitive reactance for C_3. If the output impedance matches the input impedance of the next stage, the value estimated for C_3 will suffice. For a mismatch, some compromise may have to be made to relate the value of the reactance of C_3 not only with R_L but also with the base-input resistor of the following stage.

In the previous discussions regarding calculation of the capacitance values and reactance of the coupling and bypass capacitors, the *impedance* factors that exist were not considered, since the procedures outlined are totally adequate and precise design procedures are not required. This is particularly the case where so many variables exist in the requirements for a particular stage that a slight change in gain, or a few units of deviations in component values, will not materially affect the total characteristics. If the prototype built from the design procedures does not check out to expectations, changes are then made where required to produce the desired results.

Coupling capacitor C_1, for instance, has a specific reactance which, in combination with the resistance of R_2, forms an impedance across which input voltages drop. Other capacitances, however, also influence the voltage drops, including the interelement capacitances between base and emitter, plus the capacitances existing in the circuit feeding an input signal to the C_1 and R_2 network.

To utilize all the available values of stray capacitance, interelement capacitances, and lead inductances, and to interrelate them into the calculations for component values, would create unnecessary complexities in the design procedures. These values are only needed where precise and close tolerances are mandatory.

3-14 DESIGN-RELATED EQUATIONS AND APPLICATIONS (Part 2—JFETs)

The basic circuit for a JFET signal-voltage amplifier is shown in Fig. 3-11 with the common-source configuration most widely used. As shown, the FET is an *n*-channel unit, and the component arrangement is similar to the bipolar junction transistor amplifier shown in Fig. 3-10.

As described in Sec. 1-9, the JFET *n*-channel type has a lower channel conduction for a negative-voltage gate bias, because it depletes the carriers within the normally "on" unit. For the circuit shown in Fig. 3-11 two sections relate to bias, the series resistors R_1 and R_2 plus the source resistor R_3. Resistor R_1 is connected to the supply voltage V_{DD}, which has a positive polarity in reference to the negative-polarity ground. If the positive voltage (with respect to the source element) is applied to the gate, gate current will flow and signal distortion will occur. This comes about because of the high input impedance of the FET units and, as described in Sec. 1-8, signal *voltage* changes are used at the input and not *current* changes, as with the bipolar transistor.

To establish a proper value of bias for this circuit it is necessary to adjust the amplitude of the voltage drop across R_2 so that it is lower in

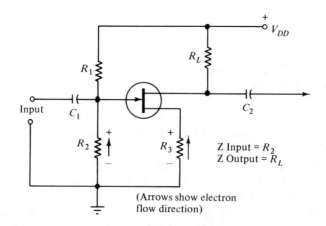

Figure 3-11 JFET *R-C* circuit

value than the voltage drop across R_3. As shown in Fig. 3-11, both these potentials are positive at the top of each resistor. For R_2, however, the voltage drop causes the gate to become more positive than the source (via the common ground connection). Similarly, the positive voltage across R_3 causes a voltage difference between gate and source; only now it would make the source more positive than the gate. Hence, if the voltage drop across R_2 is set below that which occurs across R_3, the gate will be negative *with respect to the source,* even though it is positive with respect to ground. Thus, the voltage difference between the drops is the actual bias established between gate and source elements. For zero bias, both drops are set for equal values.

Because of the extremely high input impedance of the JFET, any general value selected for R_2 automatically establishes the ohmic value of the input impedance. Thus, if the circuit furnishing the signal has an output impedance of 8 kΩ, this value resistance is selected for R_2 to provide an impedance match.

Now assume that V_{DD} is 12 V and that there is an 0.8-V drop across R_3. Bias is to be -0.5 V at gate with respect to the source. Hence, the voltage drop across R_2 must be $+0.3$ V to cancel 0.3 V across R_3 and leave 0.5 V, which makes the source element more positive than the gate by this one-half volt.

With the intended Z match and R_2 set at 8 kΩ, resistor R_1 must have a voltage drop of 11.7 V to produce 0.3 V across R_2. Thus, a 3.9-to-1 ratio of voltages must exist between the voltage across R_1 and the voltage across R_2. Thus resistor R_1 must have a value of 3.9 times 8 kΩ, or 31,200 Ω. If resistor R_1 is substantially higher in value than R_2, there will be little influence on the input impedance. If the value of R_1 is less than 10 times the R_2 value, some impedance changes will be felt, particularly if the value of R_1 drops below a ratio of 5 to 1. This comes about because the portion of R_1 tied to the V_{DD} voltage source is at signal ground by virtue of bypass capacitors and filter capacitors of the power source. Hence, it can be considered in shunt with R_2 for ac-type signals. Thus, impedance changes occur for the input of the amplifier, and R_2 may have to be raised in value so that its value combines with R_2 (in shunt) to equal the required 8 kΩ.

As with the bipolar junction transistor, the ratio of R_2 to R_3 affects stability, and a ratio of R_3 to the drain-source resistance affects gain and the degenerative effect of R_3, which rises for an increase of resistance. With R_2 having a value of 8 kΩ, one-tenth of this value would produce 800 Ω for R_3. (See also Sec. 3-13.)

As shown in Fig. 3-11, the output impedance is established by the value of R_L, and the latter is selected to provide as close an impedance match with the following stage as possible. Thus, if the following stage has an input impedance of 6 kΩ, R_L can be made the same value. If the value of the load resistor needs to be substantially higher, (above 15 or 20 kΩ of resistance) the output impedance (internal) of the selected JFET should be referenced, because if it is near the value of R_L, its shunting effect will alter the output impedance. Then the value of R_L will have to be increased so that the two values (in parallel) provide an approximate impedance match with the circuit fed by the amplifier stage.

For class A operation with a minimum of distortion for audio (or video) signals, the operating point for the JFET can be set where the voltage at the drain (V_D) is approximately one-half that of the source voltage V_{DD}. Thus, for a 6-volt drop across R_L to achieve the necessary drop, the bias is set at the point where 0.001 A circulates in the drain-source network. Hence, $E = IR = 0.001 \times 6000 = 6$ V. Now we find the voltage drop across R_3 to be $0.001 \times 800 = 0.8$ V, as mentioned earlier.

If the calculated bias is incorrect across R_3 the ratio of R_1 to R_2 must be reconsidered and changed so that the voltage drop across R_2 and that across R_3 provide the necessary difference to establish the bias needed. Some compromises may have to be made with regard to stability and gain, as with the bipolar transistor discussed earlier, to achieve optimum results.

The value of the coupling capacitor C_1 is found by the same procedures outlined for C_1 in Sec. 3-13 for the circuit of Fig. 3-10. Since R_2 is much higher here, however, a small-value coupling capacitor can be used. For one-half of 8 kΩ and at 40 Hz as used earlier, we obtain

$$C_1 = \frac{1}{6.28 \times 40 \times 4000} = 1\,\mu F$$

Thus, a 1-μF capacitor can be used, or a larger one selected to reduce additionally the series reactance offered to low-frequency signals for C_1. For each doubling of the capacitance value a corresponding halving occurs for the reactance (for a given low-frequency cutoff signal). For the coupling capacitor at the output (C_2) the same considerations prevail as were detailed in Sec. 3-13.

When the output signal is obtained from across the source resistor R_3 [refer to Fig. 2-2(c)] the input impedance equals the value of the gate resistance, but the output impedance then becomes

$$Z_{out} = \frac{R_L}{1 + g_m R_L} \tag{3-27}$$

where R_L = the source-to-ground resistor acting as the output load resistor.

Design-Related Equations and Applications / 79

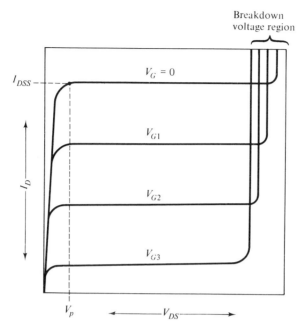

Figure 3-12 V_p and BV sectors

Specifications for FET units generally include values for I_{DSS} and V_p, and these relate to the points shown in Fig. 3-12 on the characteristic curves. Here the V_p refers to the pinch-off voltage, which occurs at the point of intersection of the zero line of V_G with the lowest V_{DS} where I_D becomes independent of V_{DS}. The I_D at this point is referenced as I_{DSS}. (See also Table 7-9.) The pinch-off voltage has also been defined as given earlier in Sec. 2-3, where it occurs at the point where the I_D drops to zero for a specific V_{DS} along the $V_G = 0$ curve.

The breakdown region occurs where the V_{DS} reaches a value that breaks down the internal insulating barriers of the FET and an excessive amount of drain current flows. The abbreviation BV is used for the breakdown voltage, though BV_{DS} can be used to designate the drain-source breakdown voltage, or BV_{DG} for the drain-to-gate breakdown voltage. (See also Fig. 2-6 for I_{DSS} and V_p points on the transfer characteristic curves, and related discussions in Sec. 2-5.)

As discussed in Sec. 2-5, the transconductance of an FET changes for different points on the transfer characteristic curve. If we solve for g_m on the $V_G = 0$ line, we can use the I_{DSS} and V_p values for finding the transconductance (g_m):

$$g_m = \frac{2I_{DSS}}{V_p} \qquad (3\text{-}28)$$

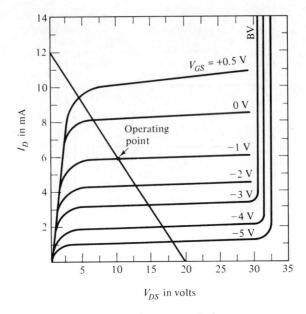

Figure 3-13 JFET gate-to-drain curves

As an example, assume that we wish to find the transconductance along the $V_G = 0$ line for the curves shown in Fig. 3-13. If $I_{DSS} = 8$ mA and $V_p = 5$ V, we obtain

$$\frac{2 \times 0.008}{5} = 0.0032 \text{ mho} = 3200 \, \mu\text{mho}$$

To solve for the load resistance R_L shown by the load line for these curves, we use Eq. (3-1):

$$\frac{20 - 0}{0.012 - 0} = 1666 \, \Omega$$

Signal gain is given by Eq. (2-3):

$$(g_m \times R_L) = 0.0032 \times 1666 = 5.3$$

This is the signal voltage gain (approximately only) for the middle-frequency response. In an audio amplifier the shunting capacitances within the transistor (plus the input capacitances of the next stage) tend to attenuate higher frequency signals. For a particular transistor two so-called *cutoff* frequencies prevail. The upper cutoff frequency occurs where the frequency response drops off 3 dB, and is sometimes referred to as the *half-power point* or *level*. For the lower frequency end the low-frequency cutoff point also occurs where the signal drops 3 dB below the average response level.

The low-frequency end is influenced by the value of coupling capacitors, and hence these should be sufficiently high in capacitance values to provide for a low reactance for the lowest frequency to be passed, as previously discussed.

3-15 DESIGN-RELATED EQUATIONS AND APPLICATIONS (Part 3—MOSFETs)

The basic characteristics and bias factors for the MOSFETs were covered in Chapter 1, Sec. 1-8. Also, MOSFET gain factors were discussed in Sec. 2-9 of Chapter 2. Both these explanations should be referenced in conjunction with the present discussions. A MOSFET resistance-capacitance coupled type of circuit is shown in Fig. 3-14; it is comparable to the bipolar and JFET circuits shown earlier in Figs. 3-10 and 3-11. For Fig. 3-14 the common-source circuit is shown with an *n*-channel enhancement MOSFET.

As with the JFET, the MOSFET units have a high input resistance, though with the insulated-gate factor the resistance input between the source and gate is infinitely high (over $10\,T\Omega$); hence $Z_{input} = R_2$ for Fig. 3-14, as was the case for the JFET discussed in Sec. 3-14. Also, as with the JFET, the output resistance of impedance is equal to R_L. (See also Sec. 2-10.)

For RF amplifiers the interelement capacitances of the transistor have a much lower reactance (X_L) than for audio signals. (See also Sec.

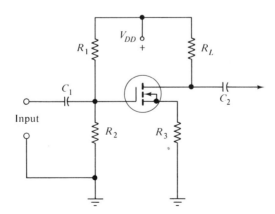

Figure 3-14 Enhancement MOSFET *n*-channel *R-C* circuit

4-2.) Hence the reactive components must be considered in shunt with the input and output impedances; consequently the following equations apply:

$$Z_{input} = \frac{R_2 X_C}{R_2 + X_C} \qquad (3\text{-}29)$$

$$Z_{output} = \frac{R_L X_C}{R_L + X_C} \qquad (3\text{-}30)$$

A typical set of characteristic curves for an enhancement MOSFET is shown in Fig. 3-15. Since no negative potentials are shown, this represents an *n*-channel type. As the gate voltages (V_{GS}) are raised, conductivity is *enhanced* and hence drain current (I_D) rises accordingly. To illustrate basic calculations, a load line has been drawn with the operating point set where the V_{GS} voltage equals that of the V_{DS} and is 7 V for both. The power-supply potential maximum applied (V_{DD}) is approximately 12.5 V as shown, where the load line intercepts the *x* axis. Thus, although the V_{DD} is not exactly twice that of the operating voltage V_D, good linearity is achieved, for an input signal swing of 2 V (from $V_G = 8$ to $V_G = 6$) produces an equal change of V_{DS} from 4 to 10 V. Thus our signal-voltage gain is

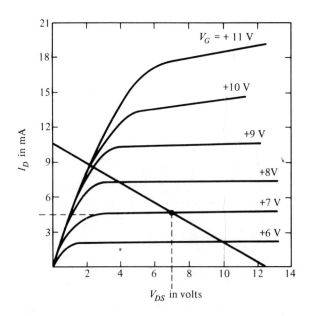

Figure 3-15 I_D versus V_{DS} curves for enhancement MOSFET

$$A_e = \tfrac{6}{2} = 3$$

We can calculate the value of R_L by using Eq. (3-1) for a dE/dI:

$$R_L = \frac{12.5}{0.0105} = 1190.5 \, \Omega$$

From Eq. (2-2) we solve for transconductance (g_m) along the operating voltage (7-V V_{GS}), where a two-volt change of V_G produces an I_D change of from 2 mA to 7.5 mA, or 5.5 mA:

$$g_m = \frac{0.0055}{2} = 0.00275 = 2750 \, \mu\text{mho}$$

We can verify the gain of 3 previously found by using Eq. (2-3) ($g_m R_L$):

$$A_e = 1190.5 \times 0.00275 = 3$$

Typical characteristic curves for a depletion-type MOSFET are shown in Fig. 3-16. The positive V_G potentials above the $V_G = 0$ can be considered as the enhancement mode, because progressively higher

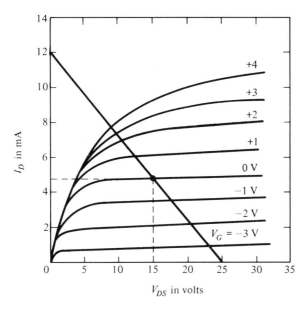

Figure 3-16 I_D versus V_{DS} curves for depletion MOSFET

positive potentials enhance conductivity and I_D rises. Below the $V_G = 0$, progressively higher negative potentials deplete the electron carriers in the channel. (See Sec. 1-9.) Note that where the gate potential is 0 current still flows. Thus the depletion type is normally in a conductive state.

For Fig. 3-16 the load line is drawn to set the operating point on the zero bias line. The slope of the load line carries it to the 25-V point along the V_{DS} axis as shown, while the other end of the load line ends at 12 mA. For a two-volt change of V_G we obtain a change along V_{DS} of approximately 12.5 V to 17.5 V, or a difference of 5 V. Hence, $\frac{5}{2} = 2.5$ for the voltage gain (A_e).

For the value of the load resistor we obtain

$$R_L = \frac{E}{I} = \frac{25}{0.012} = 2083 \, \Omega$$

The transconductance (g_m) along the V_{DS} operating point is

$$g_m = \frac{0.006 - 0.0035}{2} = 0.00125 \text{ or } 1250 \, \mu\text{mhos}$$

To prove the gain of 2.5 obtained above, we multiply the transconductance by the load resistance:

$$A_e = 2083 \times 0.00125 = 2.6$$

This 2.6 value is sufficiently close to the 2.5 originally obtained for verification, since slight differences will exist for approximate readings derived from the characteristic curves.

CHAPTER 4

RF Amplification

4-1 CIRCUIT CONSIDERATIONS

Radio-frequency amplification is a term used in relation to amplifiers that handle signals substantially above the audio range. Thus, RF amplifiers process such signals as those transmitted by AM and FM stations, those of this type obtained by receivers and processed in the tuner and IF stages, and also for the much higher frequency signals encountered in television receivers. The IF amplifiers can also be considered RF amplifiers, since they handle high-frequency signals.

While in early stages of transmitters the RF amplifiers handle a single-frequency signal (the carrier), other RF amplifiers must be capable of amplifying a group of signals having various frequencies around a central primary frequency. This is necessary because of the sideband signals generated during the modulation process.

In transmitters, the basic RF signal (*carrier*) is modified by the modulation process, so the carrier can convey the audio or video signals. The modulation process generates additional signals of various frequencies immediately above and below the carrier frequency, such signals constituting *sidebands*. Hence, an RF amplifier must be capable of amplifying to the same degree both the carrier and sideband signals generated by the modulation process. Thus, a specific bandwidth is required in the stages of the RF amplifier.

The sidebands plus the carrier signal thus occupy a specific space in the frequency spectrum and conform to the Federal Communications

Commission specifications, which also allocate the particular carrier frequency that a station must employ.

As detailed in later sections of this chapter, the ability of an amplifier to discriminate against unwanted signals is known as *selectivity*. In AM broadcasting, the selectivity of an RF amplifier is such that it ranges to approximately 10 kHz, depending on local conditions. In public-entertainment frequency modulation the bandpass required is 150 kHz, with 200 kHz allocated by the FCC. The additional two 25-kHz spectrum spans on each side of the 150-kHz bandpass are referred to as *guard bands*, since they tend to guard against overshoot of the allocated spectrum. With FM the *extent* of carrier deviation is related to the modulating signal's amplitude. Thus, high-amplitude audio modulating tones swing a carrier much more than soft tones. Thus, FM is actually a wide-band system compared to AM. Wide-band selectivity is also necessary in television transmission and reception, where the tuner amplifier must have a selectivity ranging to 6 MHz.

To obtain a required specific selectivity, an amplifier must be designed to reject unwanted stations on each side of the desired one by using tuned resonant circuits. The selectivity Q of such circuits is determined by introducing a certain amount of resistance so that the bandpass will be sufficient to accommodate the carrier plus the sideband signals. Also, as described later in this chapter, several RF stages in cascade may be used, each one tuned to a different frequency, so that the overall response coincides with that desired.

4-2 RESONANT FACTORS SUMMARY

The equation for inductive reactance is $X_L = 6.28fL$, while that for capacitive reactance is $X_C = 1/6.28fC$. Thus, as signal frequencies are raised, the reactance for an inductance increases, but that for capacitance decreases. Similarly, for a decrease of frequency, inductive reactance is reduced. but capacitive reactance is raised. These inverse effects of X_L and X_C thus indicate that for any combination of inductance and capacitance there must be a signal frequency that will cause both reactances to be equal numerically. When this occurs, the condition known as *resonance* has been achieved.

With resonance, circuit impedance is purely resistive, since the effects of the opposing reactances cancel, leaving only circuit resistance to impede current flow. With a series-resonant circuit formed by an inductor and capacitor, Z is low and I is high. With parallel resonance, Z is high and I is low, as illustrated later in this chapter.

Because $X_L = X_C$ at resonance, $X_L - X_C = 0$. Expressed with the omega sign for angular velocity, this becomes

$$\omega L = \frac{1}{\omega C} = 0 \tag{4-1}$$

From this, we get

$$\omega^2 = \frac{1}{LC} \quad \text{or} \quad 6.28 f^2 = \frac{1}{LC} \tag{4-2}$$

Hence,

$$LC = \frac{1}{\omega^2} \tag{4-3}$$

From these relationships we can procure useful formulas for finding unknown values for resonant circuit conditions. Equation (4-2), for instance, indicates that $6.28 f^2 = 1/LC$; hence,

$$f_r = \frac{1}{6.28 \sqrt{LC}} \tag{4-4}$$

where f_r is the resonant frequency
   L is in henrys
   C is in farads

The radical sign can be removed from the equation by the following conversion:

$$fr^2 = \frac{1}{4\omega^2 LC} \tag{4-5}$$

We can convert Eq. (4-4) for capacitance in microfarads (μF) by changing the 1 in the numerator to 10^3, or to picofarads (pF) by changing the 1 to 10^6. In either case, however, Eq. (4-4) indicates the resonant frequency at which the reactance values for the inductance and capacitance achieve numerical equality.

Example: If an inductor of 0.2 μH and a capacitor of 0.8 μF are placed in series, what is the resonant frequency?

Answer: The frequency is solved by the following:

$$f_r = \frac{1}{6.28 \sqrt{0.2 \times 0.8 \times 10^{-12}}} = \frac{1}{2.5 \times 10^{-16}}$$

$$= 398 \text{ kHz}$$

Hence,

$$X_L = \omega L = 6.38 \times 398{,}000 \times 0.2 \times 10^{-6} = 0.5 \, \Omega$$

and

$$X_C = \frac{1}{\omega L} = \frac{1}{6.28} \times 398{,}000 \times 0.8 \times 10^{-6} = 0.5 \, \Omega$$

In Eq. (4-4) the product of LC is the factor that determines the resonant frequency. Hence, there are a number of L and C combinations that can be used to produce the same resonant frequency so long as the LC product remains the same. With a product of 0.098, for instance, a frequency of 500 kHz is obtained, regardless of the individual values of L (in μH) and C (in μF) producing the 0.098 product. This is shown in the following example:

Example: If a 0.49-pF capacitor is in series with a 0.2-H coil, what is the resonant frequency? What is the LC product?

Answer: Converting L to microhenries and C to microfarads yields the following LC product:

$$200{,}000 \times 0.00000049 = 0.098$$

$$\text{Resonant frequency} = \frac{1}{6.28 \sqrt{0.2 \times 0.49 \times 10^{-12}}}$$

$$= 509 \text{ kHz}$$

Therefore,

$$X_L = \omega L = 639{,}304 \, \Omega$$

$$X_C = \frac{1}{\omega C} = 639{,}304 \, \Omega$$

Other values of inductance and capacitance still produce the same resonant frequency, provided that L and C values are selected to give the same 0.098 product. Such values could, for instance, be 0.196 μF for the capacitor and 0.5 μH for the inductance: 0.196 × 0.5 = 0.098. Substituting these L and C values in Eq. (4-4) again gives a frequency of 509 kHz. Now, however, the reactance values have changed from those given earlier:

$$X_L = \omega L = 1.6\ \Omega$$

$$X_C = \frac{1}{\omega C} = 1.6\ \Omega$$

Note how the originally high reactive values of 639,304 were reduced to a low 1.6 Ω while the same resonant frequency was maintained. This wide latitude in the selection of reactance values at resonance provides a considerable variation in the choice of circuit characteristics to suit particular requirements. Thus, in addition to using Eq. (4-4), it is sometimes advisable to find the LC product so that different values of L or C can be chosen if existing ones do not provide the circuit characteristics desired in terms of signal selectivity, as more fully described later.

Equation (4-5) can be converted to find unknown values of L or C at resonance. Thus, for a specific frequency and a desired capacitance, the required inductance is found by

$$L = \frac{1}{(2\pi f)^2 C} \tag{4-6}$$

A similar equation can also be used for finding the capacitance value necessary to obtain resonance for a given inductance and frequency. The equation is

$$C = \frac{1}{(2\pi f)^2 L} \tag{4-7}$$

4-3 BANDWIDTH RELATIONSHIPS

When a signal with a frequency other than that of resonance is impressed on a resonant circuit, either X_L or X_C predominates, depending on whether the signal's frequency is above or below resonance. For such a signal that does not find resonance, the circuit no longer behaves as one of pure resistance. Since the latter condition occurs only for the resonant-frequency signal, the circuit is obviously highly sensitive to such a specific frequency. Hence, the resonant circuit has the ability to *select* a signal at or around the resonant point and to discriminate against signals lying outside the resonant region, both above and below the resonant frequency.

A graph of the resonance-characteristic curve of a circuit provides a visual indication of its selectivity. For a series circuit curve, current is plotted for applied signal frequency, values of R, L, and C being kept constant. As an illustration, assume that we have a series circuit resonant at 4 MHz, with an applied signal amplitude of 0.1 V and a resistance of 20 ohms, as shown in Fig. 4-1. Current values are then read for

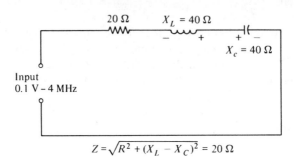

Figure 4-1   Circuit for Sec. 4-3 discussion

various frequency values of the input signal both above and below the resonant frequency. These current values are plotted along the y axis with frequency along the x axis. A few values for these conditions are given in Table 4-1 for reference.

When the foregoing values are plotted, the results are as shown in Fig. 4-2(a). Had the current calculations been extended to include a frequency span of from 62.5 kHz to 256 MHz, the graph then would have appeared as shown in (b). Hence, though both graphs plot the current for signal frequency, the degree of selectivity is not indicated. Yet, the selective characteristics of each are the same, because both are a

TABLE 4-1

Frequency in MHz	X_L in ohms	X_C in ohms	X_T	Z	Current in mA
32	320	5	$j315$	315	0.3
16	160	10	$j150$	151	0.6
8	80	20	$j60$	63	1.6
4	40	40	$j0$	20	5
2	20	80	$-j60$	63	1.6
1	10	160	$-j150$	151	0.6
0.5	5	320	$-j315$	315	0.3

plot of the same circuit, the only difference being in the range of frequencies plotted along the *x* axis. Note, also, that the curves are plotted with reference to a logarithmic scale, because the 1.6-A point on the low-frequency slope of the curve occurs at 2 MHz, and the 1.6-A point on the high-frequency slope occurs at 8 MHz. Hence, the lower-frequency 1.6-A point is 2 MHz from resonance, while the upper-frequency 1.6-A point is 4 MHz from resonance. Similarly, the lower-frequency 0.6-A point is 4 MHz from resonance, while the upper-frequency 0.6-A point is 16 MHz from resonance. If the curves were plotted for successive MHz steps, we would no longer have symmetrical slopes on each side of resonance.

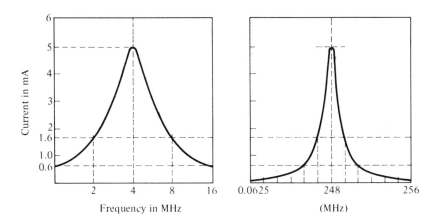

Figure 4-2 Graph differences for circuit in Fig. 4-1

In Fig. 4-2(a), our resonant frequency is 4 MHz. Obviously, however, a frequency such as 4.5 will still produce a current flow fairly high on the graph as compared to a 1-MHz or a 16-MHz signal. Similarly, a frequency of 3.5 MHz would also cause a high current flow. Thus, a resonant circuit curve indicates that a series resonant circuit will pass a *band* of signals of frequencies around the resonant frequency, but will have a diminishing pass-band characteristic for signals at greater distances from the center resonant frequency.

For graphs, the standard reference for the bandpass characteristics of a circuit is defined as the distance (horizontally) between two points on the graph, with each such point designating a signal frequency at which the amplitude of the slope is 0.707 of the peak amplitude. These points are also referred to as *half-power points*. These are illustrated in Fig. 4-3(a), where a curve is shown that is for a circuit resonant at 1 MHz. Here f_1 is the signal frequency at which the amplitude of the low-frequency slope is 0.707 of peak; f_2 shows the frequency where the amplitude of high-frequency slope is 0.707 of peak current. Hence, *bandwidth* is the distance between these points, or

$$\text{Bandwidth} = f_2 - f_1 \qquad (4\text{-}8)$$

If these points occur on the 995-kHz and 1005-kHz portions of the curve, the bandpass in kilohertz would be $1005 - 995 = 10$ kHz. Such a symmetrical frequency span on each side of f_r in Fig. 4-3(a) can be achieved by stagger-tuning and other methods described more fully in Sec. 4-18. In practical circuits the bandpass varies considerably, depending on the type of circuit required and the amount of signal frequencies necessary to transmit and receive the intelligence that is carried. Radio receivers, for instance, have a bandpass of approximately 10 kHz. For public-entertainment frequency modulation at 88 to 108 MHz, however, each FM station is allocated a 200-kHz bandwidth. In television broadcasting, a station is allocated a 6-mc bandwidth, which includes a 50-kHz bandwidth for the sound portion of the telecast. Radar, satellite, commercial, and other such services have bandwidths that differ from each other, based on their individual requirements.

The response curve for a *series*-resonant circuit can also be plotted for impedance versus frequency. Since the impedance is a function of E/I, the response curve is identical to the current curve, except that it is inverted. The *impedance* curve for the circuit graphed in (a) is shown in Fig. 4-3(b). Because the reactances cancel at series resonance, the impedance drops to its minimum values and is composed solely of resistance, as previously explained in this chapter.

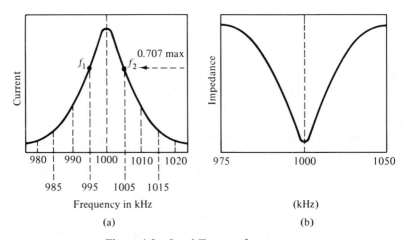

Figure 4-3 *I* and *Z* versus frequency

For a parallel-resonant circuit the impedance is high at resonance, while the current is at its lowest point. Hence, for a parallel-resonant circuit the graph in Fig. 4-3(a) would be a frequency plot against Z, while that in (b) would be a frequency plot against I. A series-resonant circuit offers a low impedance for the resonant-frequency signal; consequently, signal currents are high. The parallel-resonant circuit offers a high impedance for signals at resonance; hence signal voltage drops across the resonant circuit are high, though signal currents are low.

4-4 SELECTIVITY (Q)

When the response curve of an RF amplifier is plotted, the more narrow bandwidth denotes a greater degree of *selectivity* (symbol Q). The selectivity, as the term implies, relates to the ability of an amplifier to select the amplified signals at and near the resonant points, while discriminating against undesired frequencies lying outside a specific bandwidth.

Mention has been made in this chapter of the effect of resistance on the resonance and impedance of a series circuit. When the series resistance is increased, the bandwidth also increases, and selectivity decreases. The selectivity Q changes for R values to a considerable extent, as shown in Fig. 4-4. In (*a*), the two curves have identical inductance and capacitance values, but different resistors. For the lower value resistor of 500 ohms, the response curve is at a higher level, showing increased signal current transferred, and it is also narrower at the half-power points than the response curve obtained for a 1000-ohm series resistor. By doubling the resistance value, as was the case here, signal current decreased to one-half its original value. For signal frequencies

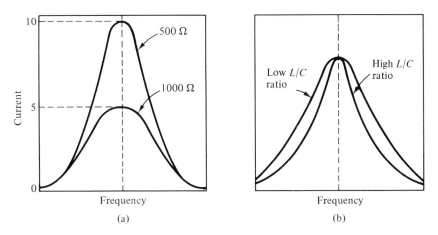

Figure 4-4 Q changes for R values

that are distant from the resonant frequency the two curves tend to blend, but the 0.707 points of the lower curve produce a much wider bandwidth and hence decrease selectivity. Thus, the broader curve is unable to discriminate against undesired frequencies to the same degree as the narrower curve.

The shape of a response curve can also be altered by changing the ratio of the inductive reactance to the capacitive reactance. With a higher L/C ratio, Q increases (the response curve narrows), as shown in Fig. 4-4(b). For these two curves, the resistance has the same value, but the *product* of LC is still the same; thus the same resonant frequency for the signal prevails.

When the inductance-capacitance ratio is altered, it follows that the ratio of inductive and capacitive reactance changes. Consequently, an equation can be formulated that takes into consideration both the X and R factors. The reactance most likely to have some resistive component is X_L, because some resistance is always present in the wire making up the coil winding.

At RF regions, the coil windings are not as numerous as at audio frequencies, but some resistive effects are still present. With air-dielectric tuning capacitors or varactor diode frequency tuning, the leakage resistance is virtually zero. Thus, coil resistance must be lumped with whatever other series resistance prevails in the circuit, so the equation is based on the ratio of the inductive reactance to the resistance. The equation is

$$Q = \frac{X_L}{R} \qquad (4\text{-}9)$$

For a series-resonant circuit the selectivity can be increased by the following: increasing the inductive reactance, decreasing the resistance, or by both increasing the inductance and decreasing resistance. When the inductance is increased to get a higher L/C ratio, the capacitance value must be decreased proportionately to maintain the same LC product.

Because the selectivity of the circuit is related to the width of the band (pass-band), the Q is related to the $f_2 - f_1$ points previously mentioned. Hence, the equation originally given [Eq. (4-8)] can also be expressed as

$$f_2 - f_1 = \frac{Rf_r}{2\pi f_r L} = \frac{Rf_r}{2\pi f_r L} = \frac{R}{2\pi L} \qquad (4\text{-}10)$$

For calculating the frequency of either f_2 or f_1, the following applies:

$$f_2 = f_r + \frac{R}{4\pi L} \tag{4-11}$$

$$f_1 = f_r - \frac{R}{4\pi L} \tag{4-12}$$

Equation (4-9) (Q equals X_L/R) can also be expressed as

$$Q = \frac{f_r}{f_2 - f_1} \tag{4-13}$$

Hence

$$f_r = Q(f_2 - f_1) \tag{4-14}$$

and

$$\text{Bandwidth} = \frac{f_r}{Q} \tag{4-15}$$

The frequency points need not be known to solve for Q, when the following equation is used:

$$Q = \frac{1}{R}\sqrt{\frac{L}{C}} \tag{4-16}$$

From the foregoing equations, resonant circuits can be designed to have the required bandpass characteristics to meet the desired requirements. A specific group of signals will be handled, while undesirable signals will be rejected. Signal transmission of radio, television, FM, and other similar services always requires sufficient bandwidth to accommodate the additional signals that carry the audio or video information. Such accompanying signals, sidebands, cluster on each side of the resonant-frequency carrier signal. Similarly, tuned resonant circuits are required at the receiver for any of these services to enable us to tune to a desired station (by changing either inductance, capacitance, or the voltage on a varactor, associated with the resonant circuit). Hence, a transmitted signal is selected as well as the associated sidebands by designing a circuit resonant to them, and by providing a bandpass that tends to reject other stations (signal frequencies) above and below the resonant bandpass. With successive amplification stages using resonant circuits, the selectivity characteristics can be increased to the level where virtually no interference from undesired signals occurs except for unusual reception conditions.

The following example utilizes equations previously given in this section to indicate specific applications:

Example: A series-resonant circuit has a capacitor of 50 pF in series with a 2.2-mH coil. The resistance of the circuit is 221 Ω. What is the bandwidth of the circuit and the Q? What is f_r? What are the frequencies at the 0.707 point of the curve?

Answer:

$$Q = \frac{1}{R}\sqrt{\frac{L}{C}} = \frac{1}{221} \times 6632 = 0.0045 \times 6632 = 30$$

$$\text{Bandwidth} = \frac{R}{6.28L} = \frac{221}{0.0138} = 16\,\text{kHz}$$

$$f_r = Q(f_2 - f_1) = 30 \times 16{,}000 = 480\,\text{kHz}$$

$$f_2 = f_r + \frac{R}{4\pi L} = 480{,}000 + 8000 = 488\,\text{kHz}$$

$$f_1 = f_r - \frac{R}{4\pi L} = 480{,}000 - 8000 = 472\,\text{kHz}$$

Knowing the frequency values, we can prove the Q value by

$$Q = \frac{f_r}{f_2 - f_1} = \frac{480}{16} = 30$$

When one is solving for circuit characteristics, the particular equations are selected that will be most suitable for the problem involved. Occasionally a different approach is preferable or may provide a more ready means for arriving at the solution. In the foregoing example, for instance, the resonant frequency could have been ascertained initially and then the reactance to resistance ratio could have been used for solving Q, as shown below:

$$f_r = \frac{1}{6.28\sqrt{LC}} = 480\,\text{kHz}$$

$$X_L = \omega L = 6632 \, \Omega$$

$$Q = \frac{X_L}{R} = \frac{6632}{221} = 30$$

Assume that in the following example a more narrow bandpass is required and hence selectivity must be increased. This could be accomplished by increasing the L/C ratio, as mentioned earlier, or by decreasing resistance. The L/C ratio, with C and L in microunits, is $0.00005 \times 2200 = 0.11$. We could double the inductance value and have the capacitance:

$$C = 0.000025 \, \mu F$$
$$L = 4400 \quad \mu H$$
$$LC = 0.11$$

Now, our bandwidth becomes

$$\text{Bandwidth} = \frac{R}{6.28L} = \frac{221}{27{,}632 \times 10^{-6}} = 8 \, \text{kHz}$$

As shown, by doubling the inductance value and halving the capacitance value, the bandpass has been narrowed. Similarly, the narrow bandpass (and consequent increase in Q and resonant current) can be accomplished by using less resistance. Retaining the original inductance value, but halving resistance, produces:

$$\text{Bandwidth} = \frac{110.5}{13{,}816 \times 10^{-6}} = 8 \, \text{kHz}$$

$$Q = \frac{X_L}{R} = \frac{6632}{110.5} = 60$$

4-5 Q FACTORS (Series and Parallel)

The opposing (inverse) relationships between series and parallel circuits must be kept in mind. With parallel circuits, the inductor may have an appreciable series resistance that must be considered. For lowering Q, however, a resistor is placed *in shunt* with the parallel inductor and capacitor to regulate bandwidth and hence selectivity. Current relationships for a parallel resonant circuit are shown in Fig. 4-5. With such a circuit, once energy has been delivered to the resonant combination of capacitance and inductance, very little additional current is drawn from the source (or generator), because the energy is exchanged between the inductor and the capacitor at a rate corresponding to the frequency of the resonant circuit. Thus, this interchange of energy (known as *flywheel effect*) will be diminished gradually because of the power consumed by any resistance in the circuit. In practical circuits the energy is constantly replenished by the amplifier and power supply and, of course, is constantly transferred to the next stage.

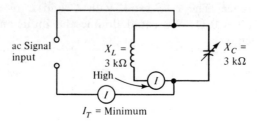

Figure 4-5 Currents in parallel resonance

If the capacitance of the parallel resonant circuit is small, it has a decreased ability to store electric energy, and the charge and discharge rates become more rapid, resulting in a higher resonant frequency for the signals. If the inductance of the resonant circuit is decreased, the resonant frequency is also increased. As with the series-resonant circuit, therefore, the frequency is determined by the L and C values. Thus, Eq. (4-4) applies here also.

As shown in Fig. 4-6, bandpass (selectivity) is again at the half-power points (0.707) on the low- and high-frequency slopes, as with the series-resonant circuit shown in Fig. 4-3. Hence, bandwidth is again between the two points $f_2 - f_1$.

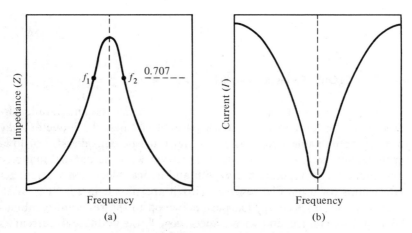

Figure 4-6 Parallel-resonant Z and I curves

When the resonant circuit is loaded, power is drawn from it, and collector current rises. The current value at resonance and under load is, however, still less than it would be if the resonant circuit were detuned.

4-6 PARALLEL RELATIONSHIPS

When parallel resistance is present, the selectivity is altered accordingly and in equation form is the reciprocal of the Eq. (4-9) for Q given earlier for the series circuit. For the parallel circuit with R in shunt the equation is

$$Q = \frac{R}{X_L} \tag{4-17}$$

If parallel resistance is zero or negligible, circuit Q is determined only by whatever resistance is present in the coil. Such inductor resistance is equivalent to series resistance (that is, as though the resistor were in series with the inductor, and both in parallel with the capacitor). Hence, the equation for the series circuit would apply to the parallel resonant circuit.

During the time that no load system is applied to the parallel-resonant circuit, and if coil resistance and shunt resistance are negligible, the following equation would apply:

$$Q = \frac{Z}{X_L} \tag{4-18}$$

The bandwidth is again related to the resonant frequency Q, and Eqs. (4-13) to (4-15) also apply. When the resistance of the wire making up the inductor is appreciable, performance and circuit Q are affected, and consideration must be given to the change which occurs. Unlike the series-resonant circuit, the circuit resistance can affect the resonant frequency in a parallel circuit. Thus, even though X_C and X_L have identical values for a given frequency, the resonant frequency may be slightly lower because of the effects of the inductor's resistance. Such resistance produces an *impedance,* since the resistance is equivalent to a series resistor, and the resistor and inductor together form Z. Thus, the shunt for the capacitor consists not only of a pure inductive reactance, but by an impedance that now affects the nullifying aspects of the two reactive currents, because it is higher in value than would be the case if the inductive reactance were considered without the resistance. If the frequency of the applied signal is decreased slightly, inductive reactance also decreases and capacitive reactance increases, thus helping to balance the reactive currents.

In most applications the difference is negligible, and a slight change in frequency is not noticed when a variable capacitor is used to tune the resonant circuit. Since, however, the resistance can be a factor in a few critical applications, the circuit resistance should be held to a minimum and larger-diameter inductor windings should be used if necessary.

At very high-frequency operation, the so-called *skin effect* present in conductors contributes to circuit resistance. As the frequencies of signals are raised, wires present a sufficient amount of inductive reactance internally (because of magnetic fields) to force current to flow on the outside (skin) of the wire. Consequently, the practice is to increase wire diameter at the higher frequencies (FM, TV, etc.) to minimize the skin-effect losses. If the increase in diameter of the wires increases capacitance effects between windings, this new factor can be reduced by spacing the wires to widen the insulation gap between them.

4-7 RF CIRCUITS

As with audio circuitry, RF circuits can be operated as class A, B, or C, depending on the requirements. As with the amplifiers discussed earlier, the bias polarity and potentials for the input system establish the type of characteristics obtained. For RF amplifiers, resonant circuits are employed because such circuits are concerned with the reception or transmission of single-frequency carrier signals plus sideband components, if present.

A typical RF amplifier using the common-emitter system is shown in Fig. 4-7. Here, a *pnp* transistor is used, though an *npn* could also be employed by reversal of the source voltage polarities. If this circuit is in a receiver, it could be the RF stage preceding the mixer oscillator section of the tuner, and the RF signal input could be from an antenna. If this circuit were used in a transmitting as a class C amplifier, the input could be from a previous stage of similar design, and the output could be to a succeeding amplifier stage or an antenna system.

As shown, the input signal is applied across the transformer primary L_1 and coupled to the secondary, L_2. The secondary has capacitor C_1 in shunt, and together they form a resonant circuit. Since the base-emitter input of the transistor may have a lower impedance than the parallel resonant circuit, L_2 is tapped from the top to the required amount to obtain an impedance match. The emitter circuit has the standard stabilization network consisting of $C_4 + R_2$. Design factors for this network are identical to those discussed for a low-frequency amplifier.

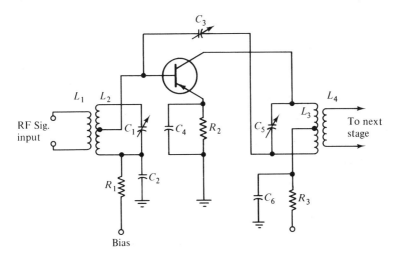

Figure 4-7 Common-emitter RF amplifier

The amplified output signal is developed across the parallel-resonant circuit consisting of C_5 and L_3 and transferred by the transformer function to the secondary L_4 as shown in Fig. 4-7. The supply potential is applied to the center tap of L_3 as shown, with a decoupling network consisting of C_6 and R_3. As with the decoupling networks discussed earlier (Sec. 3-8), the system isolates the amplifier and prevents common coupling via the power supply. The bias input network on the base side also acts as a decoupling network, since the series resistor R_1 is shunted by capacitor C_2. Both capacitors C_2 and C_6 have a low reactance for the resonant frequency signal and by having a return path to ground, these capacitors provide a closed RF circuit to the bottom of the stabilizing network in the emitter (C_4 and R_2).

Capacitor C_3 is a neutralizing capacitor, which is adjusted to minimize tendencies for this stage to go into oscillation. The factors concerning this system are discussed in the next section.

4-8 NEUTRALIZATION

When a triode transistor circuit is designed as an RF amplifier, as in Fig. 4-7, and a tuned circuit is present at both the input and output sections, the entire circuit could oscillate; that is, it could generate a signal of its own instead of simply amplifying the incoming RF signal. (This condition prevails only if the frequency of the input signal is the same as that of the output. If the circuit is used as a signal frequency doubler, with the output circuit tuned to twice the frequency of the incoming signal, oscillations would not occur, unless they were of a parasitic nature, such as would be the case for an unstable transistor operation at higher frequencies.) As discussed in Chapter 5, RF oscillators are formed by permitting circuits such as this to generate signals of their own.

Interelement capacitances in a solid-state device tend to provide coupling between the input and the output circuit, where the amplified version of the input signal is developed. Also, interelement capacitances tend to contribute to the total capacitance of the circuit that, in conjunction with the inductor, forms resonance. Hence, the interelement capacitances are a contributing factor to oscillations and as such are undesired, since they can cause interference with the input signals that are to be amplified.

When an RF amplifier such as that shown in Fig. 4-7 oscillates, it is necessary to neutralize the circuit by coupling back a portion of the signal developed at the output by use of capacitor C_3, termed a *neutralizing capacitor*. This capacitor is adjusted so that it has the same capacitance value as the capacitance within the transistor elements.

As shown in Fig. 4-7, the collector voltage supply via R_3 is to a tap on the inductor L_3; hence a signal can be obtained from the *bottom* of this inductor, which is 180 degrees out of phase with the collector signal at the top of the inductor. Thus, the capacitance of C_3 couples a portion of the signal from the collector side to the base side for neutralization of the oscillations.

For solid-state circuitry, the word *unilateralization* has been widely used as a common designation for the neutralization process. The term "neutralization" had been extensively used with vacuum-tube RF amplifiers. A unilateral circuit is considered to be one having a single-direction path for the signal to be amplified. Hence, if an input signal is applied, an amplified replica appears at the output. If, however, a signal is applied or is present at the *output, no path in reverse* to the input is present, and hence the signal is unable to reach the input side.

Unilateralization should be considered the process by which an external feedback system is designed and adjusted for canceling both *resistive* and *capacitive internal coupling* between the input and output circuitry of a transistor, plus any other reactive internal coupling. Neutralization, on the other hand, is considered to cancel only *reactive* internal coupling between output and input.

4-9 COMMON-BASE RF AMPLIFIER

As with audio amplification, the common-base system can also be employed for an RF amplifier, as shown in Fig. 4-8. Input and output transformers are again used, as with the common-emitter type of Fig. 4-7. With the circuit in Fig. 4-8, however, adjustable transformer cores are shown (tuning slugs) with the arrows between windings denoting variable cores. Such interstage tuning is often used in the intermediate-frequency (IF) amplifiers of receivers. Variable capacitor tuning could, of course, be employed as required.

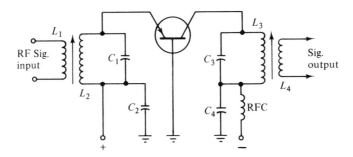

Figure 4-8 Common-base RF amplifier

As with the common-emitter circuit, the bottom of each resonant circuit is placed at ground potential, or to an RF return circuit of low reactance to ground. Thus, the input system has a closed loop around the base, emitter and tuned circuit, while the output has a closed loop between collector-tuned circuit and base.

A radio-frequency choke is used in the supply feed line at the collector instead of a series resistor. These coils are often used at radio frequencies to provide a high reactance (or impedance, if some coil resistance is present) for the resonant-frequency signals but passing the required dc. Thus, the RFC in conjunction with capacitor C_4 forms a decoupling network.

In comparison to the common-emitter amplifier, the common-base type is less affected by low-frequency oscillations, because it has less gain for frequencies substantially below the resonant frequency than the common-emitter circuitry. Also, signals with frequencies above the resonant frequency are not diminished to an appreciable amount and have an extended span before they tend to decline. This facilitates wide-band operation when such is required. Also, with the common-base RF amplifier, the output circuit is effectively isolated from the input by the grounded-base element. With the base at signal ground, emitter-base capacitances are electrically isolated from the collector-base capacitances. Hence, there is rarely any need for unilateralization measures as with the common-emitter RF amplifier.

4-10 FET RF AMPLIFIERS

As shown in Fig. 4-9, field-effect transistors are also useful for forming RF amplifier circuits. Since the impedances for the FET are much higher than for the bipolar junction transistors, the circuits in Fig. 4-9 do not tap the input and output inductors, as was the case for the circuit in Fig. 4-7.

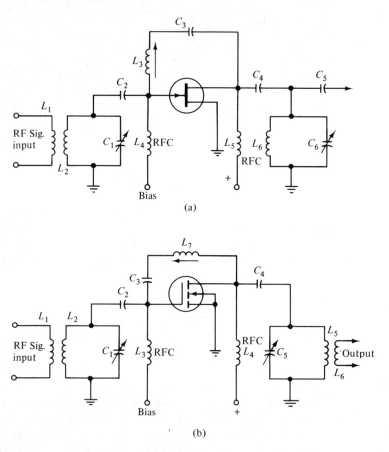

Figure 4-9 JFET and IGFET RF amplifiers

Both RF amplifiers shown in Fig. 4-9 illustrate an alternate method for applying supply potentials to the circuitry. In (a), for instance, note that at the input, inductor L_4, which is a radio-frequency choke (RFC), is in series with the bias feed line to the gate input of the JFET. Similarly, on the drain side, inductor L_5 is in series with the positive-voltage feed line as shown. Capacitors C_2 and C_4 prevent the dc supply potentials from being grounded because of the grounded resonant-circuit inductors, which would offer a short circuit for the dc.

With this type of feed, the applied potentials shunt the resonant circuits; hence power-supply current does not flow through these circuits. Such design is termed *shunt feed,* in contrast to the *series feed* shown for the amplifiers in Figs. 4-7 and 4-8. The design advantage of such a feed system is that dc potentials are not dropped across the tuning capacitors (C_1 and C_6), and this is of particular advantage when high voltages are used.

The radio-frequency choke (RFC) inductors are in series with the dc feed lines to isolate the signal from the supply system and thus minimize RF signal leakage. Since the reactance value is determined by frequency as well as inductance ($X_L = 6.28fL$), the resonant frequency of the signal being amplified must be considered and sufficient inductance must be included to provide for a reactance that can be as high as twice that of the drain-to-source impedance of the FET. If, however, the reactance is made excessively high, the additional turns of wire required for the RFC tend to increase the distributed capacitance of the inductor and thus provide an increasingly lower capacitive reactance, which will begin to shunt signals and thus defeat the purpose for using the RFC units.

If the input signal is to be amplified (and the frequency not doubled), unilateralization circuitry must be added; for the circuit in Fig. 4-9(a), this is provided by inductor L_3 and capacitor C_3. The capacitor C_3 also prevents direct coupling of the dc between the gate and drain elements. The degree of inverse feedback obtained is regulated by the values selected for the series L_3 and C_3. A variable core for the inductor permits tuning for resonance to the precise frequency required. As the inductor and capacitor are tuned closer to the resonant frequency of the signal, the impedance of the series circuit drops, and hence more feedback energy is coupled through this network to the gate input of the FET for neutralization purposes.

The output from the circuit shown in (a) is coupled to the next stage with capacitor C_5. Both capacitors C_4 and C_5 must have a low reactance for the frequency of the signals being amplified. Transformer coupling could, however, also be employed by forming a transformer with inductor L_6 to a secondary winding, as is the case for the second circuit shown in Fig. 4-9. (See also Sec. 4-17.)

In Fig. 4-9(b) is shown an *n*-channel enhancement IGFET in an RF amplifier circuit similar to the one shown in (a) for the JFET. For the system in (b) a resonant circuit (C_3 and L_7) provides the necessary coupling from output back to the input for neutralizing the oscillatory effects that may be present in similar fashion to the circuit in (a).

The process for adjusting L_7 consists of removing the input signal. When this is done, there should be no output from the amplifier unless it is oscillating, in which case it would be generating a signal of its own. A wavemeter or other signal-sensing unit is applied to the output, and L_7 is adjusted until no output signal is obtained, indicating the elimination of the self-oscillating characteristic. The input signal is then reapplied for normal amplification processes.

4-11 MODULATED CLASS C

As with audio circuitry, RF amplifiers can be operated in push-pull to increase output power and raise efficiency. Generally the push-pull system is preferred over the single-ended systems in RF amplifiers for transmitting circuitry. Such class C amplifiers initially process the carrier signals (single-frequency), which later may be amplitude-modulated (AM) by audio, video, or other low-frequency signals, as shown in Fig. 4-10.

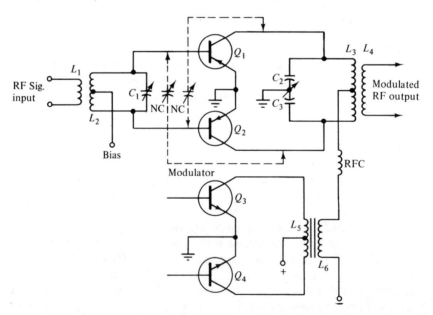

Figure 4-10 Push-pull class C amplifier (modulated)

For this circuitry the RF input signal from a previous stage of amplification, or from the generator (oscillator) of such a signal, is applied to the input resonant circuit by the transformer arrangement, as with the other RF circuits discussed herein. If the need for unilateralization arises, cross neutralization can be employed as shown by the dashed lines containing the neutralizing capacitors *NC*. Note that such cross neutralization provides feedback coupling between the collector of Q_2 and the base input of Q_1 and for the other neutralizing capacitor from the collector Q_1 to the base input of Q_2. For the output resonant circuit, split-stator capacitors are employed, wherein the rotor is at ground, minimizing shunt hazards in high-powered transmitters. The split-stator arrangement puts the variable portion at ground potential. (See Sec. 4-19, "Split-stator Tuning Design.")

Modulators for AM systems consist of high-powered high-quality audio amplifiers, usually of the push-pull type. As shown in Fig. 4-10, these feed the secondary of a special modulation transformer, with the secondary L_6 in series with the power supply feed line. Modulation is obtained by using such a transformer to add or subtract from the potentials supplied to the collectors of Q_1 or Q_2. Signals appearing across L_6 set up voltages across this secondary that aid or oppose the dc potentials as the audio signals vary from alternation to alternation. Thus, they add or subtract from the dc supply voltages, and consequently the applied collector voltages vary in relation to the amplitude changes of the modulated signals. Thus, amplitude modulation is accomplished and sideband signals are produced. A combination of sidebands plus the carrier forms a composite signal, which varies in amplitude in accordance with the modulating information.

If transistors Q_1 and Q_2 represent the final RF stages after which the signals are transmitted, the system is referred to as *high-level modulation.* This term refers to the fact that modulation occurs at the highest RF power level, regardless of the type of modulation employed. (For instance, the modulating signal could also have been applied to the base input circuits for producing amplitude modulation; this would still be high-level modulation.) If the modulated class C amplifier stage is followed by one or more class B amplifiers, the system is termed *low-level* modulation, since the modulation occurs at a level lower than that of the final output power of the last stage.

For 100 percent modulation, the signal output power of the *modulator* (Q_3 and Q_4) must be equal to one-half the dc power furnished the class C stage (the product of the class C amplifier supply voltage and the direct current flowing between collector and emitter). (The input signal to the base input of a class C amplifier transistor is termed *excitation* and must not be confused with the term *input power,* the latter relating to the $E \times I$ dc power drawn from the power supply.)

It is better not to set the modulation at 100 percent, because a continuous transmission at this level would produce amplitudes in excess of 100 percent modulation and hence would cause distortion. For practical purposes, the modulation level is kept below the 100 percent level, and modulation percentages between 60 and 80 are not unusual. The modulation percentage is defined as how much less the modulating power is than the value that is one-half of the carrier amplitude input power. In adjustment or testing procedures, a constant-amplitude modulating signal is used. The following equation applies to such a situation:

$$P_a = \frac{m^2 P_i}{2} \qquad (4\text{-}19)$$

where P_a is the modulator signal power (audio or video)
 m is the percentage of modulation (in decimal form), as 0.5 for 50 percent modulation
 P_i is the class C amplifier input dc power

Thus, if the dc input power to the class C amplifier is 800 W, for 100 percent modulation the following audio or video power is required:

$$P_a = \frac{1 \times 800}{2} = 400 \text{ W of modulating signal power}$$

For 1.5 kW at 50 percent modulation, we obtain

$$P_a = \frac{0.5 \times 1500}{2} = \frac{750}{2} = 375 \text{ W}$$

4-12 AM AND FM

The basic AM system is shown in Fig. 4-11(a) and consists of a crystal-controlled oscillator for generating the carrier, buffer class C amplifiers and class C multipliers if needed, and finally RF class C power amplifiers, which provide the highest amplification of the carrier signal. The modulator system consists of audio preamplifiers, which accept audio (or video) signals, final amplifiers and modulators, as was shown in Fig. 4-10, and then an antenna system for propagating the carrier and sidebands.

For frequency-modulating systems, several methods can be employed, and a representative system is shown in Fig. 4-11(b). Basically, all generate sideband components that, when added to the carrier, represent a constant amplitude signal, where the frequency varies at a rate determined by the frequency of the modulating signal. The *degree* of

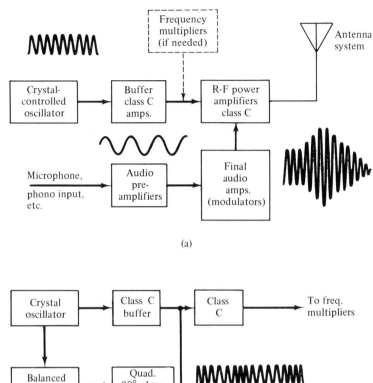

Figure 4-11 AM and FM systems

carrier shift, or the extent of such a carrier deviation on each side of center frequency, is related to the amplitude of the modulating signal. Thus, the greater the modulating signal amplitude, the greater the extent of carrier shift each side of center.

Again, a crystal oscillator is utilized, and successive class C buffer and power amplifiers to bring the carrier to the level required. A balanced modulator is used (as described more fully in Sec. 4-13), and by appropriate phase shifting the generated sidebands are combined with the carrier to form the frequency-modulation signal. An audio pre-amplifier is used to feed the audio final amplifier, the output of which is applied to the balanced modulator.

During normal frequency modulation, a number of sidebands are produced. At low-modulation levels, the number of sidebands produced in FM equals those for AM. The difference, however, is that the two sidebands produced during low levels of FM are displaced 90 degrees with respect to the carrier, as opposed to AM, wherein the two sidebands are in phase relationships to the carrier.

Thus, AM can be converted to FM by displacing the two sidebands that are produced by 90 degrees and combining them with the carrier again. The phase modulation that results can be used for frequency modulation transmission; the system is termed *indirect frequency modulation,* because the carrier is not made to shift directly by the modulating signals.

During phase modulation the deviation of the carrier relates to the frequency of the audio-modulating signal multiplied by the maximum phase shift permitted. Thus, higher frequency audio signals cause a greater swing of the carrier signal (deviation) than lower frequency signals. In frequency modulation, however, only the amplitude of the audio influences the degree of carrier deviation. To equalize the difference, a correction circuit is used, as shown in Fig. 4-11. This capacitor-resistor combination is termed a *predistorter.* The series resistor has a high resistance compared to the reactance of the shunt capacitor for the full range of audio frequency signals used. Thus, the audio signals impressed across the predistorter undergo no appreciable phase change between the signal voltage and signal current. The audio output from the predistorter is obtained from across the shunt capacitor; hence the amplitude of signals will vary for different frequencies. Thus, for increasingly higher audio-frequency signals, the audio stage receives signals of decreasing amplitude from the predistorter. Since the phase-modulation process causes a greater carrier frequency deviation for the higher frequency signals, the predistorter (having the opposite effect) nullifies the characteristics of phase modulation and converts the process to the equivalent of FM.

In television transmission, the video picture signals are amplitude-modulated and the related sound signals are frequency-modulated. More aspects of this system are covered later in this chapter.

4-13 FM BALANCED MODULATOR

A typical balanced FM modulator is shown in Fig. 4-12. This circuit serves a dual function: It modulates the carrier to create side-

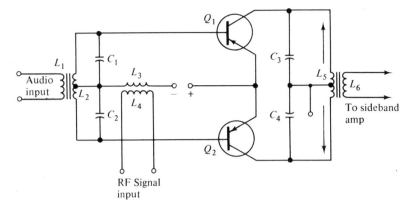

Figure 4-12 Balanced modulator

bands and also suppresses the carrier. Thus, only the sideband signals are obtained from the output of the balanced modulators.

The RF carrier signal is injected in series with the supply potential and center tap of the input transformer L_2. Thus, the RF input signal is applied *in phase* to both base elements of transistors Q_1 and Q_2. Hence, a single RF signal alternation across L_3 causes *both* base elements of the transistors to have the same forward-bias change. Consequently, if the output across L_3 opposes the negative forward bias, the drop in bias decreases current in both transistor collector-emitter circuits. Since the collectors are connected in push-pull circuitry, electron flow through each transistor is in the direction shown by the arrows at L_5. Thus, current changes on each side of the center tap of L_5 are equal in amplitude but opposite in polarity; consequently, cancellation occurs for such a current change representative of the RF signal.

The input audio signals that appear across the secondary of the input transformer (L_2) are applied to the base inputs of the two transistors. Because of the center tap, the voltage division across the secondary provides a 180-degree phase difference between the signals applied to the base inputs of the transistors. The input signals have a voltage relationship as follows for the carrier E_c and the modulating voltage E_m:

$$E_{b_1} = E_c \cos \omega t + E_m \cos \omega t$$
$$E_{b_2} = E_c \cos \omega t - E_m \cos \omega t$$

Because the audio signals cause changes in the current flow of the collectors, the carrier-frequency currents within each transistor undergo modulation. When a carrier is modulated, sidebands are produced, and these encounter resonant circuits in the output consisting of C_3 and one-half of L_5 for Q_1, and C_4 and the other half of L_5 for Q_2. These resonant circuits have a low impedance for the audio signals and minimize their appearance at the output. Because the carrier has been suppressed, only sideband signal energy is obtained from the balanced-modulator system. Capacitors C_1 and C_2 at the input have a low reactance for the RF signal energy and consequently provide coupling to the base inputs. For the audio signals appearing across L_2, however, capacitors C_1 and C_2 provide a very high reactance and thus have little shunting effect.

For proper operation, balanced circuits of this type must be symmetrical for both the upper and lower sections. The transistors should be closely matched for characteristics, gain, etc. Similarly, the inductor L_2 should be center-tapped properly, with capacitors C_1 and C_2 having the same value. Similarly, capacitors C_3 and C_4 should have equal values, and L_5 should be center-tapped correctly so that one side does not provide a greater reactance than the other.

4-14 PREEMPHASIS AND DEEMPHASIS

In public-entertainment FM (both mono and stereo) a noise-reduction process is used that involves a system known as *preemphasis* at the transmitter and *deemphasis* at the receiver. Interelement transistor noises and circuit noises are generated at a fixed amplitude level in a given system. Thus, the signal-to-noise ratio can be increased by raising the level of the signal over the constant-level noise signal. Noise levels tend to increase for higher-frequency audio signals, and if the level of the audio signals is raised at an increasing rate for the higher-frequency signals, noise reduction can be obtained. This is done by bringing the signals down to the proper amplitude at the receiver.

The Federal Communications Commission (FCC) has established regulations for noise reduction in FM. The rate of incline for the preemphasis process and the rise in amplitude of the audio frequency signals start at about 400 Hertz, and rise gradually; at 1000 Hz the increase is 1 dB; at 1500 Hz the increase is almost 2 dB. At 200 Hz, an amplitude rise of approximately 3 dB occurs, while at 2500 Hz there is almost a 4-dB rise. From this point on, the increase in preemphasis is virtually linear, reaching 8 dB at 5 kHz and 17 dB at 15 kHz.

The basic circuit for preemphasis is shown in Fig. 4-13(a). This resembles the standard capacitor-resistance-coupled circuit of an audio amplifier stage, with C_1 as the coupling capacitor and R_1 as the base resistor across which the signals develop. Instead of a large capacitance for C_1 being used to provide a low reactance, a reduced capacitance value is used to provide the necessary rise in amplification for the higher frequency audio signals. The time constant (RC) for the preemphasis network is 75 μs, as established by the FCC. That is, RC equals 75×10^{-6}. This provides optimum results without excessively increasing frequency deviations because of the increase in signal amplitude. During reception, the deemphasis is used to nullify the effects of the preemphasis. Deemphasis causes a gradual decline for the higher frequency signals, thus bringing the transmitted audio spectrum to that which entered the microphone. This normalizes the response and eliminates the harsh high-frequency audio that would occur without the deemphasis. A typical circuit is shown in Fig. 4-13(b) and consists of the series resistor R_3 plus the shunting capacitor C_3. Such a deemphasis network usually follows the FM detector and often precedes the volume control, as shown. The time constant is again 75 μs, and the higher frequency signals encounter an increasingly lower reactance for C_3, with a consequent decline in their amplitude because of the shunting effect. Capacitor C_2 is the conventional coupling capacitor having a sufficiently low reactance to provide adequate passage for the signals.

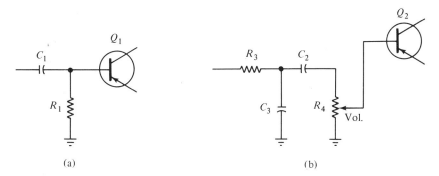

Figure 4-13 Preemphasis and deemphasis

4-15 RESONANT FILTERS

When the need arises to suppress some signals while permitting others to enter a circuit, the resonant capacitor-inductor combinations can be used for filtering purposes. Two typical examples are shown in Fig. 4-14. These are sometimes referred to as *pi* bandpass filters, because of the configuration of the three circuits used. For the filter circuit shown in (a), two parallel resonant circuits are used in conjunction with a series resonant circuit. One parallel resonant circuit may be sufficient if the pass-band requirements are not too critical.

For the resonant filter shown in (a), assume that the requirements are for passage of a 3-MHz signal, while signals above and below this frequency are kept from getting through. (Another possibility is that a group of signals clustered around the 3-MHz signal are to be passed, with a bandpass of 20 or 30 kHz.) To function as a bandpass filter, all three resonant circuits are tuned to the signal frequency that is to be passed through, as shown in (a). Thus, for the 3-MHz signal, the series-resonant circuit has a low impedance and does not hinder the transfer of this signal. For signals having frequencies above and below the resonant frequency, the series circuit provides a high impedance and attenuates such signals because of the opposition produced.

Because the parallel-resonant circuits are in shunt with the input line, they will bypass signals that find a low impedance. For the 3-MHz signal, however, a high impedance is present, and hence very little of a desired signal will be shunted. Signals not at resonance, however, find a low impedance in a parallel-resonant circuit and hence are shunted. Consequently, any significant portions of the undesired signals are attenuated at the output. Hence, the three resonant circuits combine to perform the passing and the shunting functions of the filter network.

The width of the bandpass depends on the Q of the resonant circuits. Hence, by regulating the amount of resistance as well as the L/C ratio, the width of the bandpass, and the relative admittance versus impedance for the desired and undesired signals can be regulated to meet specific requirements.

A band-stop type of filter is shown in Fig. 4-14(b). Such a filter acts in opposite fashion to the bandpass filter shown in (a). For the filter in (b), passage is permitted for all signals *above* and *below* the frequency of the resonant signal, while the latter signal is filtered out. Signals having a frequency clustered around resonant frequency would also be attenuated, depending on the width of the band-stop resonant curve.

For the circuit in (b), the parallel-resonant section is in series between the source of the signals and the output; hence a high impedance is present for the resonant-frequency signal. For other signals, the parallel circuit offers a low impedance. The two series-resonant circuits are in shunt as shown, and these have a low impedance for the signals to be shunted. Hence, the series-resonant circuits filter out the 3-MHz signals as well as those grouped around the frequency of this signal. For signals that are substantially above and below the resonant frequency, the impedance of a series-resonant circuit is high, and very little of the desired signals is shunted. The unwanted signals are bypassed by the first series-resonant circuit, and any small-amplitude signals that are left encounter a high series impedance in the parallel-resonant circuit; again, attenuation occurs. For any undesired signals that are left at the output of the parallel resonant circuit, additional attenuation is encountered at the second series-resonant circuit, which shunts the output.

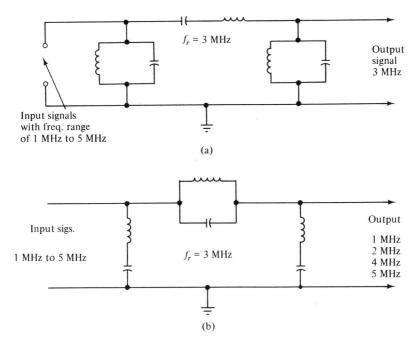

Figure 4-14 Bandpass and band-stop filters or traps

Equations given earlier can be utilized in the design of either the bandpass or bandstop filters. The equations are utilized for ascertaining characteristics or for calculating design factors. Typical problems are given below:

Example: A bandpass filter is to be designed for 321 kHz. If the inductor is to be 500 µH, what must be the value of the capacitor for the series and parallel circuits? What resistance must be placed in the series-resonant circuit to obtain a Q of 20? What is the bandwidth of the series-resonant circuit? What design factors must be observed for maximum parallel-circuit impedance?

Answer: Value of C [from Eq. (4-7)]
= 492 pF and $X_L = \omega L$
= 1008 Ω
For a Q of 20:

$$R = \frac{X_L}{Q}$$
$$= \frac{1008}{20} = 50.4 \text{ Ω}$$

Bandwidth [Eq. (4-10)]

$$= \frac{R}{6.28L} = 16.05 \text{ kHz}$$

For maximum parallel Z, inductors in parallel-resonant circuit must have minimum resistance and a Q in excess of 10.

Example: A band-stop filter has an inductance of 2815 μH in each resonant circuit, and the capacitor for each circuit has a value of 4 pF. What is the band-stop frequency? What is the Q of the series circuits if each has a 1000-Ω resistor? What is the band-stop width?

Answer: Q [from Eq. (4-9)] = 26.5
Frequency [Eq. (4-4)] = 1500 kHz
Bandwidth [Eq. (4-15)] = 56.5 kHz

4-16 TV TRAPS

Typical applications of the type of resonant filters discussed in Sec. 4-15 are the resonant circuit traps used in television receivers to filter out adjacent channel interference, as well as interference from the

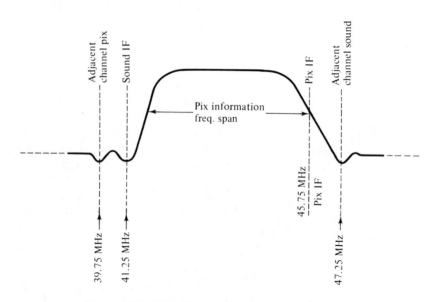

Figure 4-15 Television receiver IF response curve

sound-IF signal. The traps cause perceptible dips in the video IF response curve, as shown in Fig. 4-15, and hence attenuate such undesired signals and minimize the interference patterns that could result from the reception of such signals.

As shown in Fig. 4-15, one trap has a filter action for 39.75 MHz, which attenuates the signals received from an adjacent channel picture carrier. Another trap is used to prevent the 41.25-MHz signal from causing interference at the picture tube. Another trap reduces the interference from the adjacent channel sound, and has a frequency of 47.25 MHz. (Another trap is the 4.5 MHz needed to keep this sound-IF signal from reaching the picture tube. This is the signal obtained from mixing the 45.75-MHz picture IF with the 41.25-MHz sound IF, to produce a new sound IF of 4.5 MHz. The distribution of these signals is shown later in this section.)

As shown in Fig. 4-16, series-resonant traps are employed for shunting the undesired signals. The inductors have variable cores so that each series-resonant circuit can be tuned properly for a maximum shunting effect and the least interference. These traps need not precede the first IF amplifier as shown, but could be separated and dispersed in the other IF circuitry.

In Fig. 4-16, capacitor C_4 prevents the forward bias applied to the base of Q_1 from being applied to the tuner circuitry. Inductor L_4 in combination with C_4 forms a series-resonant circuit for the IF frequency obtained from the tuner. In the heterodyning process the dif-

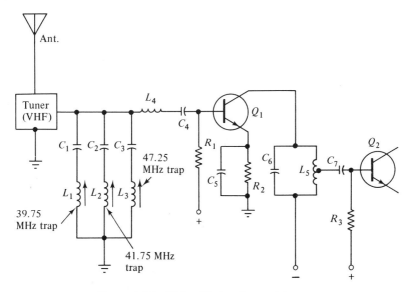

Figure 4-16 Video IF circuitry and traps

ference-frequency signals are obtained, as well as the carriers used for the mixing process. Consequently, the series-resonant circuit filters out the outwanted signals, passing only the IF signals to transistor Q_1.

Additional traps are shown in Fig. 4-17, where the video amplifiers are shown that follow the video detector. These circuits amplify the picture signal sufficiently for application to the picture tube.

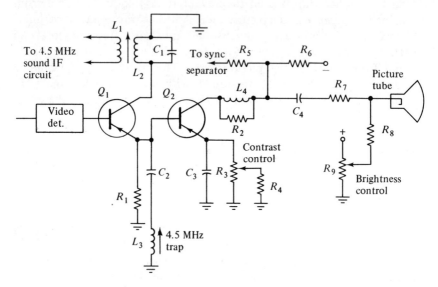

Figure 4-17   Video amplifier and traps

Transistor Q_1 has a dual output, one at the collector and another at the emitter. The collector output develops across the parallel-resonant circuit of L_2 and C_1, which is resonant to 4.5 MHz. The transformer coupling between L_2 and L_1 transfers this 4.5-MHz signal to the separate sound IF circuit for demodulating of the audio and subsequent application to the speaker. This 4.5-MHz signal would cause interference if it arrived at the picture tube; consequently, a series-resonant circuit shunts the output at the emitter of Q_1, and consists of C_2 and L_3. This low-impedance series-resonant circuit shunts the 4.5-MHz signal, and exact tuning is provided by the variable core of the inductor.

The detected picture signal that develops across R_1 is also applied to the base input of transistor Q_2, which is the final video amplifier. These video amplifiers are, in essence, high-quality audio amplifiers, capable of amplification to the 4-MHz region. The bandpass is extended by using what are termed *peaking coils*. Inductor L_4 is one such coil, and in conjunction with capacitor C_4 is broadly resonant from about 3.5 to 4 MHz. Inductor L_4 is shunted by resistor R_2 to extend the bandpass to prevent too sharp a resonant response and consequent ringing. When such circuits ring, they tend to interchange energy and to oscillate at very high frequencies. Ringing in such a circuit can cause several repeat lines extending to the right of televised scenes.

On occasion, another peaking coil may be in series with the load resistor (R_6). Such an inductor is made resonant with shunt capacitance, and the peaking coil in parallel with the shunt capacitance forms a parallel resonant circuit of high impedance. If the tuning of the parallel peaking coil and the shunt capacitance is at a bandpass range around 3.5 to 4 MHz, it offers a high impedance for such signals, and minimizes the shunting effect that would occur if only the shunt capacitance were present.

By varying the resistance of R_3 in the emitter of Q_2, the gain of the video signals can be altered, and hence contrast is regulated. Resistor R_4 prevents the complete shorting out of the emitter system. The brightness of the picture tube is regulated by variable resistor R_9, as shown. As the cathode of the picture tube is made more positive with respect to the grid, the control grid becomes more negative, and brilliancy is decreased, since less electron flow occurs between the cathode and the picture-tube face. As the resistance of R_9 is adjusted toward the ground level, negative bias to the grid is reduced, and electron flow increases from the cathode. Consequently, picture-tube brilliancy increases. Resistor R_8 has the same function as R_4; it prevents a complete grounding of the cathode for a ground setting of resistor R_9.

4-17 COIL-COUPLING FACTORS

When successive resonant circuits are coupled to each other by a transformer arrangement of the inductors, the bandpass characteristics of the individual resonant circuits are altered because of the influence of the coupled circuits. Thus, for circuits such as those shown earlier in Figs. 4-7 and 4-8, the degree of coupling (how close the primary and secondary inductors are placed with respect to the interaction of their individual fields) not only influences the gain, but also the bandpass, as shown in Fig. 4-18.

With loose coupling, the secondary current has a low amplitude and is sharply peaked, as shown. As the primary and secondary coils are brought into closer proximity, more of the load resistance of the circuit that is coupled by the coils is reflected back into the primary system. The current in the secondary peaks at a higher level, and the bandpass broadens, as shown in Fig. 4-18. When the coupling is increased to the point where the resistance reflected into the primary equals the pri-

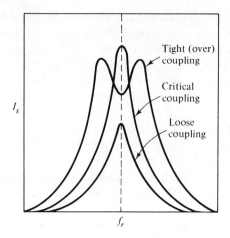

Figure 4-18 Curves for various couplings

mary resistance (an impedance match), the degree of coupling is referred to as *critical*. Now the current circulating in the secondary reaches its maximum value for the resonant frequency of the signal involved.

If the coupling is increased additionally, it is known as *tight coupling* or *overcoupling*. The response curve now undergoes a radical change, with a dip forming around the resonant frequency sector, as shown in Fig. 4-18. Thus, the gain for the resonant frequency signal is decreased and new resonant peaks now develop on each side of the center resonant-frequency point. Such coupling is often utilized for obtaining a broad-band response.

When inductors are coupled, another factor, the *mutual inductance*, must be considered. This is discussed in the next section.

4-18 MUTUAL INDUCTANCE

When coils are placed into close proximity so that their magnetic fields interact, the overall inductance has the same characteristics as those of inductances placed in series-aiding (that is, the individual coils are wound in the same direction, so that their magnetic fields aid each other rather than oppose each other). When a transformer arrangement is utilized, the RF energy is transferred by virtue of the inductance characteristics only, even though a resonant-circuit shunt capacitor is present.

The added value of inductance that results when two coils are coupled is called *mutual inductance*, having the symbol M. When an ac signal of one ampere flows in the primary, it induces one volt across the secondary, and the two inductances then have a mutual inductance of one henry. When all the magnetic lines of force of the primary coil cut the secondary coil, the equation for mutual inductance is

$$M = \sqrt{L_1 L_2} \tag{4-20}$$

where M = mutual inductance in henrys
 L_1 = inductance of primary in henrys
 L_2 = inductance of secondary in henrys

Usually, with resonant circuit coupling in RF amplifiers, the inductors are not brought so close that all the magnetic lines of force of one inductor cut the other. Thus, the mutual-inductance equation is modified to take into consideration the coupling known as the *coefficient of coupling* (indicated by the symbol k). The coefficient of coupling represents the *percentage* of coupling. Hence, a k of 0.5 indicates that only the half of the lines of force of the primary inductor intercept the secondary winding, and the coefficient of coupling is actually 50 percent. The k is added to Eq. (4-20) to form

$$M = k\sqrt{L_1 L_2} \tag{4-21}$$

Example: A transformer primary has an inductance of 3.2 μH and a secondary inductance of 0.2 μH, with a coefficient of coupling of 0.6. What is the mutual inductance?

Answer: $M = 0.6\sqrt{3.2 \times 0.2}$
  $= 0.6\sqrt{0.64} = 0.8 \times 0.6 = 0.48\ \mu$H

In a series circuit, the total inductance (L_T) is:

$L_T = L_1 + L_2 + 2M$

In the example given above, the total inductance is

$3.2 + 0.2 + (2 \times 0.48) = 4.36\ \mu$H

The foregoing value of 4 μH is of academic interest only, since the individual inductances plus their shunting capacitor make up the individual resonant circuits. If the mutual inductance and individual inductances are known, k is found by using the following equation:

$$k = \frac{M}{L_1 L_2} \qquad (4\text{-}22)$$

As discussed in Sec. 4-4, bandpass is established by the Q designed into the system. For critical coupling, the coefficient (k_c) is related to the individual circuit Q of the primary and secondary by the equation

$$k_c = \frac{1}{\sqrt{Q_p Q_s}} \qquad (4\text{-}23)$$

where k_c = coefficient of critical coupling
Q_p = Q of primary winding
Q_s = Q of secondary winding

In television and frequency-modulation systems the number of sidebands necessitates a wide bandpass. For such operation circuit selectivity is altered during the design process of the resonant circuits to procure the required bandpass width. Although overcoupling increases bandpass, the dip at the resonant frequency, as well as the two peaks above and below resonance, provides an uneven frequency response to the various signals clustered around the resonant-frequency signal. A response curve with a substantially flat top can be obtained by using two or more amplifier stages with different degrees of coupling for each. The first amplifier could have tight coupling and the second

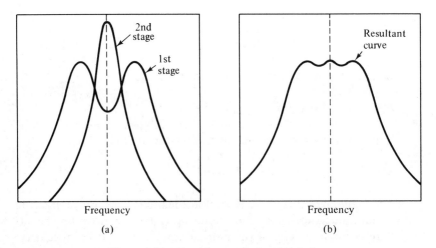

Figure 4-19 Double-tuned characteristics

amplifier critical coupling, as shown in Fig. 4-19(a). The result is a combination of the two curves, which produces the substantially flat top response curve shown in (b). With careful design and adjustment of the coupling, as well as regulation of the amplitude of the second stage, the dips in the flat top are negligible if the total curve amplitude is fairly high, as shown in (b).

Similar results of wide-band operation with a flat-topped response curve can also be obtained by using a system known as *stagger tuning*. This system is designed so that each successive stage is tuned to a slightly different frequency around the resonant curve. Again, the combined amplification response results in a single wide-band curve of the type shown in Fig. 4-19(b). The response curve depicted could be the result of three stages, with one tuned slightly below resonance, the other at resonance, and the next above resonance. The stagger-tuning method permits the design of a symmetrical response curve where f_1 is as far from f_r at the low-frequency portion of the curve as f_2 is from f_r on the high-frequency side. The degree of selectivity desired can be critically set by the particular resonance points of the individual stages.

4-19 SPLIT-STATOR TUNING DESIGN

Earlier, in Sec. 4-11, the modulated class C amplifier was discussed, and the accompanying illustration in Fig. 4-10 showed a split-stator capacitor arrangement in the output resonant circuit. Such a system has several advantages over the single-capacitor resonant circuit (see Fig. 4-7) in which the collector inductor portion of the resonant circuit is center-tapped to split the inductor in two so that an inverse signal can be obtained and applied to the base input via a variable neutralizing capacitor (C_3). A portion of this circuit has been reproduced in Fig. 4-20(a) for comparison purposes with the split-stator type shown in (b) (and also in Fig. 4-10).

In transmitting, where class B and class C amplifiers of this type are utilized, high-RF power circulates in the components, and a single-capacitor type of the kind shown in Fig. 4-20(a) has the rotary shaft of C_2 above ground for RF and dc potentials. Thus, when the necessity arises for tuning the circuit to resonance by using C_2, there is a danger of electric shock on contact. This disadvantage is eliminated by the split-stator circuit shown in (b), which provides for a balanced condition while also having a common shaft for the rotor that may be placed directly at ground as shown. Thus, the danger of RF or dc shock is eliminated. The inductor L_2 is center-tapped, providing an RF potential at each end with respect to ground. This provides for an ideal system for feeding a balanced load.

124 / RF Amplification

For the split-stator circuit system, the inductor size must be increased and the capacitor values decreased in comparison to the single-capacitor circuit in (a) for the same power dissipation and drive-signal characteristics.

Of significance are the factors illustrated in (c) of Fig. 4-20. Note that zero RF signals prevail at the ground (node) points at the rotor of the split-stator capacitor, and at the center tap of inductor L_2. Also note that no bypass capacitor is used for the power-supply connection to the center of inductor L_2. A radio frequency choke minimizes signal losses to the common power-supply system.

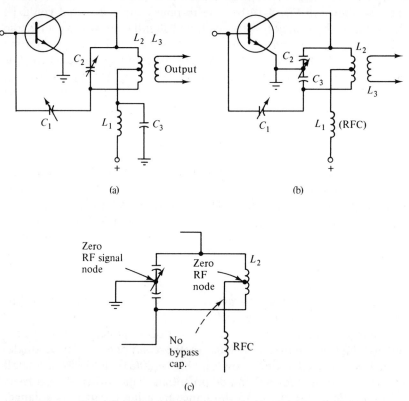

Figure 4-20 Single-tuning capacitor versus split-stator capacitor

As shown in Fig. 4-20(b), the emitter of the transistor is connected to the junction of C_2 and C_3 (common ground), and the collector is connected to the junction of C_2 and inductor L_2. Consequently, capacitor C_3 is actually in series with the inductance, as illustrated for the equivalent circuit shown in Fig. 4-21(a). Since the reactance of a capacitor opposes the reactance of an inductor, the reactance (X_C) of C_3 cancels a portion of the inductive reactance (X_{L_2}) of the coil, and somewhere along the inductance L_2 an RF zero voltage (node) occurs. Because this zero-potential point corresponds to RF ground, the circuit is effectively split in half, as shown in Fig. 4-20(c).

As the node point thus established in the inductor may be difficult to localize by trying to tap the inductor at exact center, it is left undisturbed. The bypass capacitor is *not* used, because it would place the point at which it is attached at RF ground, which may differ from the node automatically established by the circuit design. To prevent the power supply (feeding dc to the system) from causing the point of attachment to be placed at direct ground via the power-supply filter capacitors, the dc is applied to inductance L_2 through an RF choke for isolation purposes, as mentioned earlier. If the power-supply tap at the center of the inductor is bypassed, efficiency will be impaired, because a nodal equilibrium established automatically will be disturbed.

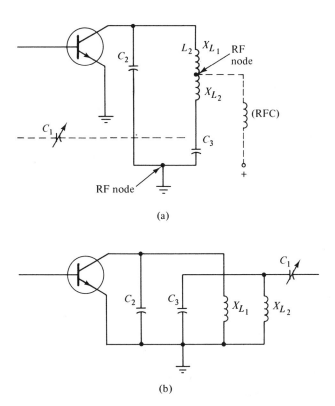

Figure 4-21 Equivalent circuitry

The equivalent circuit in Fig. 4-21(a) shows that the transistor is actually connected to one-half of the resonant circuit. This is more clearly understood by reference to the redrawn version shown in (b). Compare this with the markings shown in (a): C_1 is a neutralizing capacitor; C_2 is one-half of the split-stator system, and is connected from collector to ground. Inductor L_2 is also placed at the collector, and half of it reaches the RF node point, which is signal ground. Capacitor C_3 is connected from true ground to the inductor (X_{L_2}) and the RF node, which is equivalent to the RF ground.

In design, several factors must be considered when one is establishing the proper L to C ratio for resonance at a particular frequency. It must be remembered that parallel-resonant circuits have a high impedance when no load is connected to them, but as soon as a load (coupled to a subsequent stage, or an antenna, etc.) is applied, the load imposed on the circuit will decrease overall impedance to a comparatively low value. This is indicated by the rise in collector current when the load is coupled to a resonant circuit of this type.

To reduce undesirable formation of harmonic signals in a class C amplifier, the circuit must have a proper Q value *under load conditions*. If the Q value is too low, it will broaden the bandpass response and decrease gain; hence the circuit will tend to accept rather than reject harmonic signals. Also, selectivity as well as signal power will decrease with the lower Q. On the other hand, an excessively high Q (under load conditions) causes a high circulating current in the resonant circuit with greater resultant losses. Too high a Q also manifests itself as higher than normal current ratings when the circuit is tuned to resonance without a load. With no load applied, the impedance is high, and current drops to a very low value unless the circuit is not operating properly. An extremely high Q usually results in considerable RF radiation (representing a loss of signal energy from the resonant circuit).

For optimum results, the circuit Q should be approximately 12 when the resonant circuit is fully loaded and properly tuned to the resonant frequency, regardless of the type of transmission (telegraphy or telephone).

When the transistor is across one-half of the split-stator resonant circuit, *that half* should still have the same Q as a single-capacitor-coil combination. Proper Q is obtained by suitable selection of L/C ratio in relation to the correct load resistance for the collector in usage.

With the split-stator system, the total current drawn from the power supply is still the same with a given power input to the transistor. However, current divides through each portion of the circuit, because the section connected to the transistor couples power into the split section. Thus, each portion now obtains one-half of the total power. As current divides but total power remains the same, it is evident that an equivalent parallel resistance is now applied to the output system represented by the transistor collector-emitter. This means that each portion of the circuit must now have a load resistance equal to twice that of a conventional single-capacitor circuit, because cutting the current in half will double the value of R_L. It is assumed, of course, that no change is made in the load circuit and that it consumes the same power as would be the case for a single-capacitor resonant circuit.

Because the coupled load increases the electric current, it follows that for class C operation, with a given power input, the resistance reflected by the secondary or coupling network is proportional to the ratio of dc collector voltage to dc collector current. This may be expressed as follows:

$$R_L = \frac{E_c}{I_c} \tag{4-24}$$

For a given Q, the tank capacitance that is required for a given frequency will be inversely proportional to the reflected resistance presented by the load. This capacitance is, then, inversely proportional to the E_c/I_c ratio, and for an unsplit resonant circuit may be calculated as follows:

$$C = \frac{300 Q I_c}{f E_c} \tag{4-25}$$

where Q = a constant of approximately 12
I_c = total dc collector current in mA
f = frequency in MHz
E_c = dc collector voltage
C = total capacitance in pF across resonant-circuit inductor

The foregoing equation solves for the operating capacitance of the variable capacitor and does not refer to the maximum or minimum tuning range. A range that tunes around the calculated value must be chosen, to provide the tuning limits desired.

When current is decreased by one-half in the part of the circuit connected to the transistor, it is the same as increasing E/I ratio. As the ratio has actually been doubled owing to the halving of the current, and as C is inversely proportional to the E/I ratio, C is cut in half for the portion of the tank circuit connected to the collector. To bring this section of circuit back into resonance again, the coil inductance must be doubled, because C has been halved. As the inductance in the second portion of the circuit must also be doubled, there is now a total inductance that is *four times* the inductance encountered in single-capacitor resonant circuits. (This is obtained by *doubling* total turns, which quadruples total inductance of the resonant circuit inductor.)

Power for the entire resonant circuit is not increased when a changeover is made from a single to a split-stator capacitor system. Instead the prevailing power only divides for each section. The inductor is hot for RF at each end, and current must still be supplied to each section; but it is now only one-half the value for each. Total current, however, is the same as before, because the same total resistance is encountered. Even though the resistance in each half of the inductor adds to twice that of an ordinary resonant circuit, the two coil sections are virtually in parallel, thus paralleling the reflected resistances and bringing the R_L to the total value as before.

Solving for one capacitance in split-stator circuitry on the basis of one-half the circulating current in the capacitance equation previously given indicates that the new capacitance is one-half of that which would be used for a single-capacitor circuit. Thus, *each capacitor* in the split-stator system is now *one-half* of what the single-capacitance circuit would be. The total capacitance of the split-stator section is one-quarter, and the total inductance of the accompanying coil is four times that of the single-capacitor circuit.

The split-stator resonant-circuit system is also of advantage for push-pull operation, because it provides an ideal method for circuit division when neutralization is required. As shown in Fig. 4-22, cross-neutralization can be used without upsetting the individual resonant-circuit arrangements. Capacitor C_3 is connected from the bottom of the resonant circuit to the base of transistor Q_1, and capacitor C_4 is connected to the base of Q_2, from the top of the resonant circuit as shown. Again, no bypass capacitor is used where the power system taps the resonant-circuit inductor that feeds the collectors of the transistors. (Even though a node is automatically established, the tap should be placed as near the center of the inductor as possible, because the distributed capacitance of the RF choke may still have some influence in upsetting the null point established by the circuit.)

As a split-stator circuit has the same relationships whether used in a single-ended or push-pull arrangement, all the foregoing factors apply for the push-pull circuit shown in Fig. 4-22. If a push-pull stage had the same power-input relationships as a single-capacitor system, the value of the total split-stator capacitance would be one-quarter that of the capacitance in the single-capacitor circuit. In new split-stator designs,

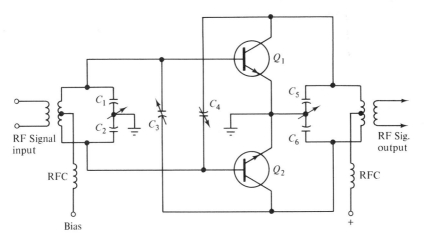

Figure 4-22 Push-pull RF amplifier

total C is calculated by using the equation given earlier [Eq. (4-7)] making each capacitor of the split-stator *one-half* the calculated value given by the equation.

When one is computing the proper capacitance and inductance values for resonance, it is convenient to consider the two sections of the split-stator system as a series circuit. Thus, the total capacitance is now established as one-fourth the value given by the equation. Once the capacitance has been determined for the desired operating frequency, the value of the total inductance (the entire coil) that will produce resonance with the split-stator capacitor can be determined by the following equation:

$$L = \frac{(159,223)^2}{f^2 C} \tag{4-26}$$

4-20 PIX IF (MOSFET)

Two stages of intermediate-frequency amplifiers for television receivers are shown in Fig. 4-23. Such stages amplify the picture and sound carriers obtained from the mixer stage of the tuner until sufficient signal amplitude is obtained for detection purposes. Usually a third IF stage is employed before the detector, with circuitry similar to that shown in Fig. 4-23. Terminal-enhancement FETs are used, and one gate for each transistor is used for AGC bias control, while the IF input signals are applied to the other gate terminal as shown.

At the tuner, the respective sound carrier and picture carrier signals are heterodyned with the signal generated by the oscillator in the tuner,

130 / RF Amplification

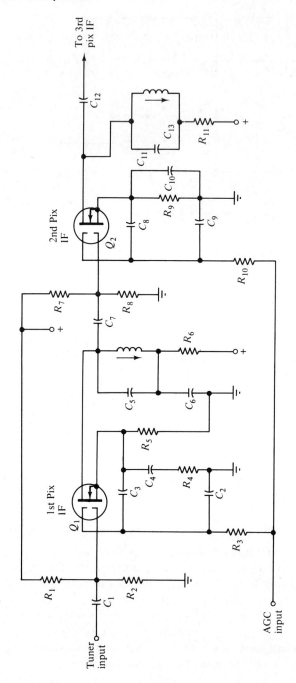

Figure 4-23 Pix IF with terminal-enhancement FETs

and the resultant intermediate-frequency (IF) signals are applied to the gate input of Q_1 via capacitor C_1. Prior to this stage, resonant-circuit traps are employed for attenuating such signals that may cause interference. (See also Sec. 4-16, Fig. 4-16.)

Resistors R_1 and R_2 as well as R_7 and R_8 form a voltage-divider network across the power source, and a tap at the center applies the necessary bias for Q_1 and Q_2. The AGC input is applied to the respective gates, resistors R_3 and R_{10} being used as shown, and filter networks are used so that the AGC signals are dc without variations caused by unwanted signals. Thus, when different stations are tuned in and the carrier signal strengths are different from those previously handled, the AGC system provides a bias change for the IF stages to compensate for the difference and hence hold the signal at a preset level.

The amplified output IF signals are obtained from the drain element and coupled to a resonant circuit composed of C_5 and the shunting inductor. The latter has a variable core so that stages can be tuned properly for maximum signal transfer. Resistor R_6 and capacitor C_6 form a decoupling network for isolation of the amplifier circuitry from the common power source. The amplified signal is then impressed on the next gate input of the succeeding stage, where similar amplifying characteristics prevail as for the first stage.

4-21 VARACTOR TUNING

Use of varactor diodes for tuning purposes is illustrated in Fig. 4-24. (See also Figs. 1-3 and 1-4.) In such diodes the interelement capacitances of the depletion region exhibit nonlinear characteristics when junction voltages are changed progressively from low to high potentials. When reverse bias is applied to the varactor diode, it becomes equivalent to a capacitor, and the value of the capacitance varies as the voltage on the unit is changed. Thus, such units can expedite tuning in radio and television receivers by replacing the variable capacitors used for tuning purposes.

As shown in Fig. 4-24, one varactor is placed across each coil for the tuner, consisting of the RF amplifier, which handles the incoming signal from the antenna, the oscillator, which generates a different frequency signal, and the mixer, which combines (heterodynes) the RF and oscillator signals to produce an intermediate-frequency (IF) signal. Each diode receives a voltage via resistor R_2 obtained from a particular variable resistor. Thus, resistor R_3 is adjusted to apply a voltage that will tune to a particular station. Next, resistor R_4 is adjusted to obtain resonance at the tuner for another station. Resistor R_5 is tuned for still

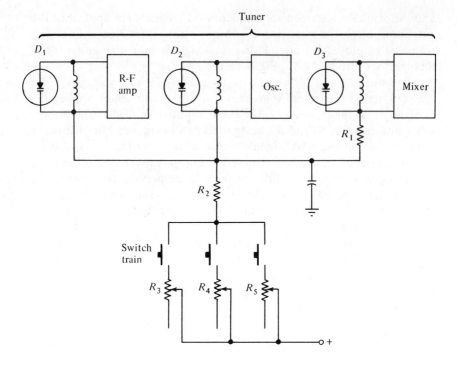

Figure 4-24 Varactor tuning

another station, and as many variable resistors are used as necessary to select the number of stations that are to be tuned by pushbutton control.

As shown, a switch train is utilized; when one button is depressed to select a station, another button that might be depressed is released automatically. A rotary switch could also be used to engage the variable arm of a particular selector resistor as the shaft is turned.

In design, diodes and resistors are selected to provide sufficient latitude to tune in the station within range of the spectrum selected for tuning. Usually any specific variable resistor is capable of tuning through the entire range. (For television, channels 2 through 13 would necessitate 12 switches for the VHF region.) For UHF (channels 14 to 83) there would also be required as many variable resistors (plus extra varactors for the UHF tuner) as needed for the stations to be selected. Since so many are involved, some manufacturers provide only a reduced number of selections, such as a total of 16, with VHF and UHF stations intermixed as desired.

4-22 IF AMPLIFIER FILTERS

In the design of high-quality receivers special features are often incorporated into the intermediate-frequency amplifier stages to improve performance and reliability, and to minimize the drift of circuitry tuned to a fixed frequency. Generally the stability of the resonant circuits is excellent, particularly in wide-band systems where slight drifting from frequency as components change may not be noticeable in performance. Also, at higher frequencies (such as in television reception or frequency modulation) the number of wire turns in inductors is low, and wire sizes are usually larger than at lower frequency operation to reduce skin-effect losses. The fewer turns and rigid characteristics of the larger-diameter wire coils tends to stabilize the tuning.

As an additional refinement, however, the IF transformers are replaced with fixed-frequency crystals or ceramic filters to help maintain a precise bandpass characteristic. These units, because of the accuracy achieved, also provide excellent rejection of undesired signals and thus enhance the general performance of the IF amplifier stages. Sometimes the filters are used in conjunction with tuned circuits, and in some instances one stage may have tuned circuitry and another only the filter sections. A typical system of this type is shown in Fig. 4-25. Here two ceramic filters are used (CF_1 and CF_2) between IF stages.

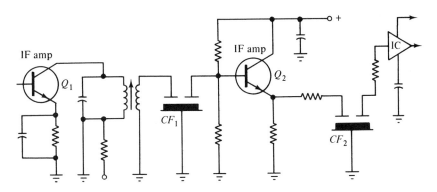

Figure 4-25 Ceramic filters (CF) in IF

The crystal or ceramic filters are not necessarily resonant to exactly the standard IF frequencies used in receivers without such filters. In the FM band from 88 to 108 MHz the intermediate frequency has been standardized at 10.7 MHz. With the crystal or ceramic filters, however, the exact intermediate frequency may be 10.6 or 10.75, and in some instances may reach 10.8 MHz. Thus, if a particular receiver uses several filters, they are matched to produce a particular bandpass, and the mixer-oscillator circuitry of the tuner must have a *difference* frequency signal output that produces the particular intermediate-frequency signal for which the filters are designed. Thus, if the ceramic filters shown in Fig. 4-25 are matched and produce a bandpass of 10.6 MHz, the tuning core between the transformer between Q_1 and Q_2 must be adjusted for maximum bandpass at the particular frequency set by the ceramic filters. Similarly, the tuner must track signal frequencies properly to produce an output signal (IF) of 10.6 MHz.

4-23 HI-FI TERMS AND RATINGS

Various terms and ratings based on acceptable standards are employed by manufacturers to list the degree by which various characteristics approach or exceed a desired value. Universal agreement on acceptable rating levels among equipment manufacturers and governmental regulatory agencies has not always been easy to achieve, and differences of opinion persist. General standards have been achieved in a number of areas, however, but the manner of testing differs among manufacturers, and the tendency of manufacturers to embellish the ratings of their own products is always a factor that must be considered. The following lists common terms and defines them in general for reference during design procedures.

Harmonic and IM Distortion. Harmonic distortion and intermodulation (IM) distortion are given in percentages related to the total signal amplitude. Harmonic distortion is tested by comparing the waveshape of the output signal with that of the input signal, with the latter set at 1 mV. Often manufacturers rate both the harmonic and IM distortion as total harmonic distortion (THD) and may give the percentage for several signal frequencies, or the total frequency span for the receiver (such as 0.2 percent for 20 Hz to 20 kHz). In most modern good-quality receivers the distortion is kept well below the point where even an experienced listener is aware of any.

When one is making comparisons of distortion ratings, the lower figure denotes the preferred rating. Thus, a rating of 0.2 percent at 1 kHz is preferred over 0.5 percent.

Total harmonic distortion ratings should be based on the full frequency response of the amplifier, because the tendency to distortion increases at the extremes of the frequency-response span (both at the low and high ends).

IHF Sensitivity. The letter combination IHF is derived from the initials of the Institute of High Fidelity; this sensitivity rating relates to the *tuner* characteristics of the radio receiver. It indicates the μV of signal required at the input terminals of the tuner (antenna terminals) to produce an audio output of 30 dB greater than the general background noise. Some manufacturers use a 50-dB reference as being more realistic. The standard has also been referred to as the μV needed at the tuner to produce an output containing no more than three percent of distortion and noises.

The FM tuner IHF sensitivity is better for the lower figure. Thus, a 1.8-μV sensitivity is superior to a 1.9-μV or a 2-μV rating, and both of these are much better than a 2.3-μV rating. Stereo sensitivity is poorer than mono because of the greater signal strength needed to produce equivalent noise-free reception compared to mono.

Signal-to-noise Ratio (S/N). This expression refers to the degree by which a receiver is capable of raising the signal amplitude above the general residual noise level. The higher the S/N ratio, the better. Hence, a 70-dB rating is superior to one of 60 or 65 dB.

Capture Ratio. This expression refers to the tuner's ability to pull in (capture) a station to which it is tuned while rejecting a weaker station at or near the same frequency. Thus, the capture ratio defines that characteristic of an FM tuner that rejects an undesired station having a specific degree of dB signal level below the desired station. Thus, a 1.5-dB capture-ratio rating means that the tuner will not pass a station through it that is 1.5 dB lower in signal strength than the one to which the radio is tuned. Hence, the lower dB rating is the preferred one. A capture ratio of 1 dB provides a rejection system superior to that of another receiver rated at 2 or 3 dB.

Selectivity. Selectivity relates to the steepness of the bandpass characteristics of the tuner and IF stages. (See Secs. 4-3 through 4-5.) The higher the selectivity ratings for a receiver, the better the bandpass characteristics. Thus, a selectivity rating of 85 dB is much superior to one of 40 or 50 dB.

Stereo Separation. This rating (in decibels) refers to the ability of a receiver, stereo amplifier, or stereo phonograph cartridge to minimize the leakage of signals from one stereo channel to the other. Generally, a 20-dB rating ensures adequate stereo separation; the manufacturer's specifications are usually based on a test performed with a 1-kHz signal. The higher the dB rating, the better the separation. Hence, a 30-dB rating indicates a superior characteristic over one having a 15- or 20-dB rating.

Power Output. This rating refers to the signal power available to drive loudspeakers. Signal-power output depends on the amplitude of the input signal and the ability of the circuits to raise such signals to the levels required to produce a given signal-power output.

Power output ratings for amplifiers have been given in rms EIA and IHF. (These letters refer to *Electronic Industries Association* and *Institute of High Fidelity*.) The rms (root-mean-square) values are the effective values of voltage, current, or power obtained by multiplying the peak value of an ac signal by 0.707. It is that power that is equivalent to the dc value required to perform the same work as the ac value. In testing for this signal power output from amplifiers a single-frequency signal input is used, and the output is measured for rms voltage across a given value of load resistance ($P = E^2/R$).

For stereo, power rating usually refers to the signal power (wattage) available per channel. Thus, a rating of 50-W rms per channel provides a total power output of 100 W for a constant-E input signal and controls set for maximum output. Speaker ohmic values of 4- or 8-Ω load conditions are usually specified by the manufacturer for a given power output rating.

Fifty-dB Quieting. This is the standard accepted generally to indicate the point of noise suppression required for high-fidelity performance. With noise suppressed to this level, the reception is sufficiently noise-free to make listening enjoyable. For stereo reception more μV of signal must be used to achieve the 50-dB quieting than for mono.

Frequency Response. This term relates to the range of audio signals passed through the amplifiers in terms of a flat response with minor variations in dB. An acceptable range would be from 20 Hz to 15 kHz (general FM broadcast limit) or to a higher value such as 20 kHz. Decibel ratings are comparisons, and a doubling (or halving) of signal power represents a change of 3 dB. A *signal* voltage doubling represents 6 dB, while a drop in signal voltage is expressed as −6 dB. For a 1- or 2-dB change, many manufacturers consider the response substantially flat.

AM Suppression (or Rejection). This term relates to an FM receiver's ability to reject amplitude modulation. Ratings are given in decibels; the higher the dB rating, the better the rejection ability. Thus, a 50-dB rating is superior to a 40-dB rating.

Image Rejection. This relates to a tuner's ability to reject image-frequency signals (heterodyning of the local oscillator with undesired signals). Other unwanted signals are usually included in this rating, such as spurious signals that may enter the system. The higher the dB rating, the better the rejection characteristics. A rating of 100 dB is thus superior to a rating of 90 dB.

Damping Factor. This relates to the loudspeaker's ability to suppress motion quickly once the actuating signal has abruptly declined to zero. A factor of 50 is superior to one of 10 or 12. Good damping reduces the "hangover" effect, where tones tend to die off slowly with low damping factors.

Tracking Force. This is a term relating to phonograph needle and cartridge characteristics. It is the grams of pressure exerted by the needle on the record during play. A lower tracking force produces less record wear; hence a needle exerting 1-gram pressure is superior to a needle having a tracking force of 3 grams. For a tracking force of less than 1.5 grams the tone arm and balance mechanism must be mechanically designed for free movement vertically and horizontally and preferably must have an antiskate device. Also, the cartridge must have good compliance for the lower tracking force operation.

Antiskate. This term refers to the process of balancing out the lateral force applied to a phonograph needle (and its tone arm) as the grooves move the assembly toward the record center during play. The device consists of a mechanical arrangement that provides a counter force that is adjustable for the particular tracking force used. The antiskate device minimizes *skating* (*sliding*) of the needle over the record surface. Proper turntable level (both for its width and depth) also aids in minimizing needle skating.

Compliance. This term also relates to phonograph needle-cartridge operation. It refers to the ability of the phonograph needle to *comply* with the variations of the record grooves; the greater the compliance, the better the reproduction of various frequency signals. Compliance can be considered as the ability of the needle to move freely from side to side within the record groove. A cartridge rating of 35×10^{-6} has excellent compliance and one of 15×10^{-6} is good, while any lower rating usually is found in the lower priced units.

CHAPTER 5
Oscillator Circuits

5-1 CRYSTAL OSCILLATOR

When a piezoelectric quartz crystal is ground to a specific thickness, it becomes a transducer which, if voltage is applied across it, distorts the crystal shape and hence converts electric energy to mechanical energy. If an ac signal of proper frequency is applied, the crystal vibrates (oscillates) at that specific frequency. Similarly, under pressure or strain, the crystal also generates a voltage. Consequently, the frequency-stabilizing characteristics of quartz crystals are employed in oscillator circuits where requirements are to hold the generated signal as close to a specific frequency as possible.

In transmitting systems, where the carrier frequency must be held precisely at resonance, the crystal is placed in a small enclosure and subjected to a controlled temperature. Such an enclosure (termed a *crystal oven*) uses a resistance-strip heating element to generate the required temperature. With most crystal types, a definite frequency change occurs for temperature variations.

The crystal is mounted between two metal plates, which form the contacting elements for the crystal circuit; the symbol is as shown in Fig. 5-1(a). The crystal can be considered equivalent to a resonant circuit, because the mass that determines vibration can be likened to the inductance, while the capacitance is contributed by the holding plates plus the mechanical compliance of the quartz structure. The mechanical friction set up during vibration is equivalent to the resistive component of a resonant circuit. Because of this, the frequency of an oscillator us-

Crystal Oscillator / 139

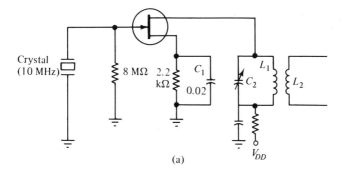

(a)

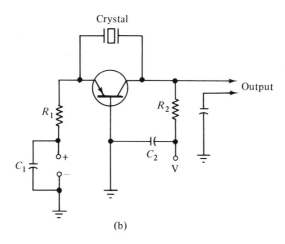

(b)

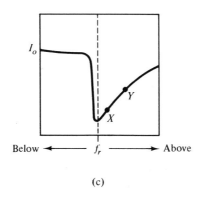

(c)

Figure 5-1 Crystal oscillators

ing a crystal can be changed to a limited degree by introducing components of inductance, capacitance, or both across the crystal. Often a small variable capacitor shunts the crystal for critical tuning purposes.

For the circuit in (a), the crystal can be considered equivalent to the resonant circuit shown at the drain output of the FET. For the circuit shown, 10-MHz operation is obtained. Component values are given for reference purposes, though these change for various types of FET units.

Figure 5-1(b) shows a *pnp* transistor with a shunting quartz crystal between emitter and collector. This is a grounded-base circuit, and there is no phase reversal between the signal developed at the collector side compared to that at the emitter. Because signal currents in the emitter circuitry are in phase with those in the collector, the only requirement for oscillations is to provide coupling between emitter and collector. The crystal slab provides such coupling and forms a signal feedback loop to establish the rate of oscillation. Since the circuit is self-sustaining, no additional resonant circuit is needed, though one could be used to form a transformer for output coupling, as shown in (a).

Best stability for the crystal oscillator is operation on the slope of the curve above the resonant frequency, shown in (c). If operation is set between points X and Y, stability will be maintained. If the circuit is tuned too close to the resonant frequency, any temperature changes can cause operation to shift below the resonant frequency point, and the circuit would go out of oscillation.

In view of the foregoing, the resonant circuit in (a) composed of C_2 and L_1 should be calculated on the basis of the operating frequencies slightly above resonant frequency. Since, of course, capacitor C_2 is variable, compensation can be made for best stability.

The reason for operation above the resonant frequency comes about because an oscillatory circuit, such as a combination of inductance and capacitance or a quartz crystal, becomes a pure resistance at resonance, since X_L is equal in amplitude to X_C, but opposite in polarity. If the resonant circuit at the output is also tuned to the exact frequency, resistance again prevails, and the entire oscillator circuit (in terms of lumped inductances and capacitances) would be unable to oscillate, because the required L and C effects are missing for energy-exchange conditions. Hence, it is necessary to tune the output resonant circuit slightly above the resonant frequency. When this is done, the variable capacitor C_2 is adjusted to decrease capacitance. Now, when the resonant circuit is tuned above resonance in this manner, it is still being pulsed by the crystal at the lower frequency rate. This means that the capacitive reactance is increased while the inductive reactance is decreased. With an inductive reactance decrease, there is a rise in signal-current flow through the inductance, causing an inductive characteristic in the parallel-resonant circuit. Now the inductance, in conjunction with lumped circuit capacitances, will permit the complete oscillatory circuit to function properly.

The resonant frequency of a quartz crystal is related to its thickness and the axis along which the crystal slab is cut. The following equation applies:

$$f_r = \frac{k}{t} \tag{5-1}$$

where k is a constant for type cut
f_r is the resonant frequency in kHz
t is thickness in inches

For the foregoing equation, if the frequency is taken in MHz, the thickness will be in thousandths of an inch. For an X-cut crystal, k reaches a maximum value of 112.6, while the so-called AT cut yields a k value of no more than 66.2. The Y cut provides a maximum cut of 77.0, and a BT cut has a maximum k of 100.78.

The following variation of the equation may be found useful:

$$t = \frac{k}{f} \tag{5-2}$$

5-2 VARIABLE-FREQUENCY OSCILLATOR

Where precise frequency control is not essential, the quartz crystal can be dispensed with, and inductors and capacitors can be used for forming the required feedback and oscillations. Typical is the circuit shown in Fig. 5-2(a). Here, an *npn* transistor is used in what is termed a *Hartley oscillator*. As shown, the resonant circuit inductor L_1 has a tap that causes an effective division of this inductor: one section coupled to the base via capacitor C_1, and the other coupled to the collector via C_3. As shown, the tap is at ground. Since inductor L_1 forms part of the circuitry of the base as well as the collector, the amplified energy appearing at the collector is coupled inductively by virtue of the common L_1 inductor to the base circuit, thus providing feedback to sustain oscillations.

Resistor R_1 furnishes the forward bias for the transistor, while the collector potential has a radio-frequency choke in series to minimize signal losses to the power supply circuitry.

Instead of a tap on the inductor L_1, the tap can be placed at the junction of two capacitors, as shown in Fig. 5-2(b). Performance is

142 / Oscillator Circuits

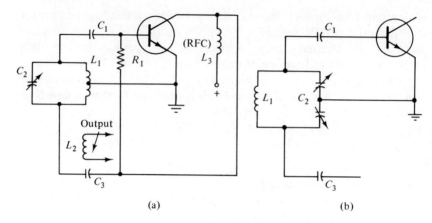

Figure 5-2 Variable-frequency oscillator

similar to that in (a), but the circuit is now termed a *Colpitts oscillator*. As with the Hartley oscillator in (a), the division of a resonant circuit provides coupling of the amplified output energy to the input circuit to create and maintain oscillations.

5-3 TUNER OSCILLATOR

An oscillator can also be formed by utilizing a special coil, which is coupled inductively to the output resonant circuit. This coil then provides a feedback path to produce and sustain oscillations. Typical of such an oscillator is that shown in Fig. 5-3, where it is used in the combination mixer-oscillator circuit of a superheterodyne radio receiver. For the circuit in Fig. 5-3, the feedback coil consists of L_2, which couples the amplified signal energy in the collector to the oscillator resonant-circuit inductance L_3. The coupling is *polarized* so that positive (regenerative) feedback occurs. The generated RF signal in the resonant circuit composed of L_3 and capacitors C_4 and C_5 is coupled back to the emitter circuit via capacitor C_3. Here, the transistor also acts as a mixer and is widely used in superheterodyne receivers, including FM and TV. The incoming RF-modulated carrier picked up by the antenna section is heterodyned (mixed) with the signal generated by the oscillator portion of the mixer transistor. Sum and difference signals are produced, and the difference signals form the intermediate frequency signal (IF), which is fed to the detector system. The emitter is also referred to as a *converter circuit*, because it "converts" the incoming signal and the oscillator signal to a *difference-frequency signal*.

Tuning capacitors consist of C_1 and C_4, which are linked together by a common rotor shaft so that both can be tuned simultaneously.

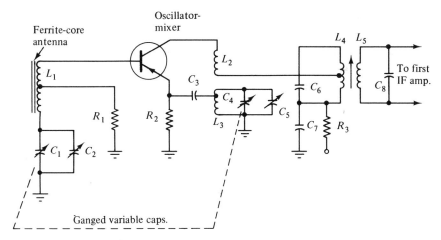

Figure 5-3 Tuner oscillator (and mixer)

Capacitors C_2 and C_5 are small "trimmer" capacitors, which are adjusted so that the two resonant circuits will track properly during tuning (that is, both maintain the exact frequency for various positions of the tuner dial). However, since C_2 and C_5 have very small values compared to C_1 and C_4, the trimmer effect is more pronounced for the higher frequency stations, where capacitors C_1 and C_4 are unmeshed and hence also have a lower capacitance.

Inductor L_2 also couples the IF signals to a tap on inductor L_4. Here a resonant circuit is formed by L_4 and C_6, and thus signals of the desired frequency (IF) develop the highest amplitude signal voltage, while undesired signals above and below resonance are rejected. A variable core between L_4 and L_5 is used for alignment and tuning of the input and output sections. Capacitor C_1 in conjunction with R_2 forms a decoupling network of the type described in Sec. 3-8.

5-4 3.58-MHz OSCILLATOR

Another application in which a crystal-controlled oscillator is very useful is in a color television receiver, where it generates a 3.58-MHz signal. Such a signal generation is necessary in a color television receiver, because only the sideband signals of color information are transmitted because of the necessity for conservation of spectrum space. Consequently, a carrier must be generated in the receiver and mixed with the sideband components for proper detection of the color signals. Not only is a crystal-controlled oscillator utilized, but it is kept precisely on frequency by special signals being transmitted by the station to which the receiver is tuned. Such signals are periodic bursts of 8

144 / Oscillator Circuits

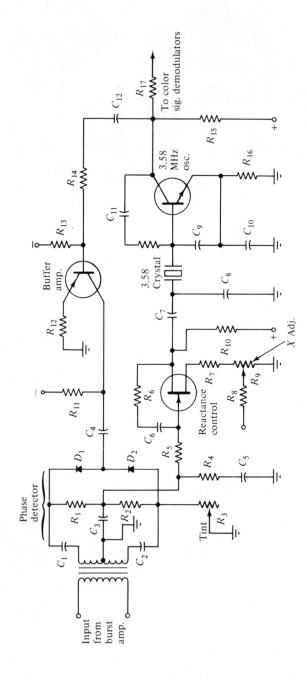

Figure 5-4 Phase-locked loop (carrier oscillator circuitry)

cycles having a frequency of 3.58 MHz. A typical circuit, shown in Fig. 5-4, comprises a phase detector, a buffer amplifier, a reactance-control circuit, and the 3.58-MHz oscillator.

The reference signal received is termed a *burst,* since it comprises only 8 or 9 cycles and is transmitted in conjunction with the horizontal sync pulses. For proper color rendition, the oscillator must be locked into synchronize with the *subcarrier* burst signal (so called because the *primary* carrier is that carrying black and white information). Even a slight drift (where *phase* differences begin to occur) must be corrected automatically. To produce such rigid control of the crystal oscillator, the crystal is ground accurately, and the frequency generated is sampled by the phase detector and compared with the incoming burst signal. If the oscillator begins to drift, a correction voltage is produced that is fed by the input of a reactance-control circuit. This circuit (using an FET in Fig. 5-4) alters the subcarrier oscillator frequency to the precise degree necessary for proper synchronization. A complete loop is formed from the phase detector through the crystal oscillator through the reactance control and back to the phase detector. Thus, this system can be considered a *phase-locked loop.*

As shown in Fig. 5-4, the incoming signals containing the burst information are applied to the phase detector input transformer, and hence appear across the secondary winding. The polarity of the signal appearing at the top of the secondary is opposite to that at the bottom, and the center-tapped ground connection splits the transformer into two sections. Capacitors C_1 and C_2 form series-resonant circuits with the secondary inductors, and the low impedance thus provided channels the signals to the respective diodes as shown. At the same time, the input burst signals are compared with the 3.58-MHz oscillator signals, which are coupled via capacitor C_{12} and resistor R_{14} to the base input of the buffer amplifier. The output from this amplifier (the collector) is coupled via C_4 to the junction of phase detector diodes D_1 and D_2.

If the burst input signals and those from the oscillator are in synchronization, no voltage is generated, and hence none is fed to the gate of the reactance-control circuit. If a phase difference prevails, however, the *difference* voltage results, and this is impressed on the reactance control FET.

The reactance-control FET circuit obtains its input signal from the junction of R_1 and R_2. The energy is then coupled to the gate input via R_5. This circuit simulates a reactive characteristic wherein its drain current lags the subcarrier oscillator voltage; thus, it acts as an inductance, and, therefore, has an inductive reactance. (See Sec. 5-5.)

The reactance FET is coupled to the oscillator by C_7 and the oscillator signal also appears between the drain and source elements of the

FET and hence across the RC network composed of R_6, C_6, R_5, R_4, and C_5. The reactance of C_6 in design is made approximately ten times that of the FET gate input impedance. When this is done, the RC network has a voltage lag (because current leads in a network that is predominately capacitive). Consequently, the signal at the gate of the FET *lags the oscillator signal.* The signal at the gate input of the FET now causes drain current to lag; hence the oscillator signal developed in the drain of the reactance FET has lagging-current characteristics compared to the signal obtained from the oscillator. Thus, the reactance control circuit simulates an inductance and hence exhibits an inductive reactance.

Because the output from the reactance-control FET circuit is coupled to the 3.58-MHz oscillator's input circuit, the reactance circuit contributes to the combined inductance and capacitance that make up the characteristics of the oscillator, including the crystal itself. Hence, any change of reactance that is coupled to the oscillator affects frequency. Even though the oscillator is controlled by the crystal, a sufficient deviation is possible to maintain precise lock-in with the 3.58-MHz burst reference signal.

Since $X_L = E/I$, any change of source-drain current in the reactance-control FET alters the amount of simulated reactance. Therefore, any change in the applied potential to the gate input changes the reactance, because it causes a change of source-drain current through the FET. Thus, if a voltage change at the gate input increases current flow through the FET, a decrease of inductive reactance occurs, equivalent to a decrease in inductance. This resultant inductance decrease across the oscillator raises the frequency of the oscillator slightly. For a decrease in the control FET current, the simulated inductance increases, causing a lowering of oscillator frequency. In actual practice, however, the circuit corrects for phase differences, so that before any appreciable frequency difference can occur, the correction is anticipated. Hence, the circuit can be considered to be locking in the phase relationships between the incoming burst reference and the 3.58-MHz oscillator output signal.

In the design of such circuitry, matched units must be used in the phase detector to maintain a good electrical balance between C_1 and C_2 as well as the two diodes. Also, the resistors R_1 and R_2 must have equal values. For the particular circuit shown, R_3 regulates the balance of the phase detector, and hence affects the tint characteristics of the color produced on the screen. Capacitor C_8 must have a sufficiently low reactance to bypass harmonics of the crystal that may upset circuit function. The reactance must not be too low, however, or it will attenuate some of the desired 3.58-MHz signal generated. If the output from the crystal oscillator has insufficient amplitude, good control cannot be maintained by the phase detector, nor will sufficient signals be available at the color-signal demodulators for proper color rendition on the screen.

5-5 REACTANCE CIRCUITRY

Basic reactance circuits are shown in Fig. 5-5. In (a) an FET is used, and the *p*-channel can be changed to an *n*-channel with appropriate bias polarity reversals. Similarly, an *npn* could be used for the transistor circuit shown in (c), again with proper polarity changes of bias. Similarly, the circuit shown in (a) can be rewired to conform to that in (c), and the transistor reactance circuit can be changed so that it has the reactance characteristics of that in (a).

The functional components making up the reactance characteristic in (a) are resistor R_1 and capacitor C_1. (The capacitor C_2 is a large-value unit, ranging up to 0.5 μF, for blocking the dc of the drain and

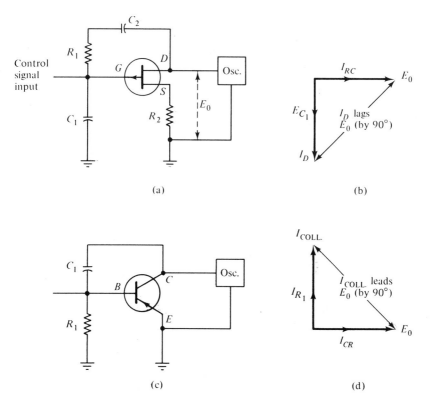

Figure 5-5 Reactance circuitry

thus preventing its being shorted to the gate terminal. For the frequency used, the reactance of C_2 is too low a value to contribute anything to the reactive function of the circuitry.)

Resistor R_1 is selected so that it has a resistance approximately ten times the *reactance* of C_1. Thus, R_1 may have a resistance of approximately 100 kΩ, while C_1 has a reactance of around 10 kΩ. The oscillator's resonant circuit is coupled across the drain and source (or across the collector and emitter) as shown. Hence, the oscillator is coupled across the $R_1 C_1$ network in (a), since C_2 has a low reactance. [Similarly, the oscillator in (c) has its voltage impressed across the network C_1 and R_1.]

The phase relationships that are established are shown in Fig. 5-5(b). In this vector diagram the oscillator voltage is represented as E_o and is designated by the horizontal line. Because R_1 is ten times as high in resistance as the reactance of C_1, the RC network is primarily resistive, so the current flow created by E_o will be virtually in phase with the voltage. Thus, the current through the RC network (I_{RC}) is also indicated by a horizontal line in (b). The signal voltage applied to the gate of the reactance FET, however, is obtained across C_1 only. Also, in a capacitor, voltage lags current by 90 degrees. Hence, the gate signal lags the RC network current by 90 degrees, as shown on the vector diagram in (b). The grounded source circuit (as with the grounded emitter) has a phase reversal between input and output, but this refers to signal voltage in and signal voltage out. For these circuits, the collector and drain *currents* are in phase with the signal voltages applied at the input base or gate. For Fig. 5-5(b), then, the drain current I_D for the vector diagram is drawn along the vertical line, to show the *in-phase* condition with respect to E_{c_1}. Obviously, then, the reactance transistor output current lags the oscillator voltage by 90 degrees. A lagging current (or a leading voltage) indicates an inductive reactance; thus, the reactance circuit shown in (a) behaves as an inductance. Since this inductance shunts the oscillator resonant circuit, the reactance transistor influences the reactance of the oscillator.

Because the circuit is reactive, any change in the drain current in (a) [or in the collector current in (c)] will influence the reactance, since $X = E/I$. Thus, the reactance can be changed by altering the forward bias applied to the input circuit.

The inductance value formed by the reactance circuit depends on the transconductance of the FET (g_m), as well as on the carrier frequency generated by the oscillator. Inductance can, therefore, be found by the equation

$$L = \frac{10}{6.28 \, f g_m} \qquad (5\text{-}3)$$

The reactance circuit using the transistor in Fig. 5-5(c) functions in a fashion similar to that discussed for (a). For (c), however, the capacitor is between the collector and base and the resistor is connected from input to ground as shown. The capacitance value of C_1 is selected so that its reactance is approximately ten times the ohmic value of R_1. Hence, current for the C_1R_1 network *leads* the signal voltage obtained from the oscillator. Since, however, the signal applied to the base of the transistor is obtained from across R_1 only, the voltage at the base will be in phase with the C_1R_1 network current, because there is no phase shift across a pure resistance. Because the signal *current* through the transistor is in phase with the voltage at the input (base), the collector current leads the oscillator signal by 90 degrees, as shown in Fig. 5-5(d). Because a capacitor produces a leading current, the circuit in (c) thus simulates a capacitor, and hence a capacitive reactance.

5-6 RELAXATION OSCILLATOR

Oscillators that do not employ resonant-frequency circuits are also available and find wide applications as low-frequency oscillators in television receivers, bias oscillators in tape recorders, and other applications including test equipment. Such oscillators are known as *relaxation* types. The frequency of the signal is related to the values of circuit resistance and capacitance.

The resonant-type oscillators produce signals that are pure sine waves. The relaxation oscillators, however, produce waveshapes that are pulses, squarewaves, or modifications of these types. The particular waveshapes available lend themselves readily to modification—to trigger other circuits, or to serve as frequency bases for sweep waveforms, as discussed more fully later. The additional components needed for such waveshape modification include discharge circuitry, differentiators, and others, as subsequently detailed.

The relaxation oscillators have characteristics permitting them to be easily synchronized by an external signal. Thus, exact frequency lock-in of such RC oscillators is used with such applications as vertical and horizontal sweep signal synchronization in television, pulse rate timing in radar, and control circuitry in industry. The two basic types of relaxation oscillators are the blocking oscillator and the multivibrator, which are described in the next two sections.

5-7 BLOCKING OSCILLATOR

A typical blocking oscillator is shown in Fig. 5-6(a). A *pnp* transistor is shown, though an *npn* or an FET could also be used by applying proper-polarity supply voltages. For the type shown in (a), regenerative

150 / Oscillator Circuits

feedback is used to initiate and sustain oscillations, so it is necessary to invert the phase of the signal that appears at the collector. For the circuit shown, the transformer secondary winding is used to feed a signal back to the base of the transistor in phase with the signal that exists there.

The signal fed back has sufficient amplitude to drive the transistor beyond the cutoff region for a time interval depending on the time constant (RC) of the circuit. A charge appears across the series capacitor to the base that opposes the normal forward bias (negative base with respect to a positive emitter). This charge gradually leaks off and conduction again occurs, repeating the cycle.

Almost any general-purpose low-voltage transistor will function in such a circuit. Because the frequency of the generating signals depends on the circuit capacitance and resistance, the internal capacitances of the transistor plus the shunt capacitances of the transformer also have a bearing on the final frequency. Similarly, the impedance of the transformer windings contributes to differences. Hence, in design practices the exact frequency of the signals generated is difficult to predict mathematically, and some experimentation is required. The transformer can be a step-down type, though a general impedance match between the output and the input circuits is preferable. With a mismatch, however, the circuit will still function reliably. The frequency can be changed by adjustment of the variable resistor in the base circuit, or by changing the capacitance of the base capacitor. Similarly, the primary L_1 of the transformer can be shunted by a capacitor for altering frequency. If oscillations do not occur, the L_2 terminals may have to be reversed to ensure regeneration, because in one position degenerative feedback would occur.

5-8 MULTIVIBRATOR

Figure 5-6(b) shows the multivibrator type of relaxation oscillator. Here, a pair of *npn* transistors is used. Instead of using a transformer, as in (a), another transistor is used for feedback purposes and phase inversion.

For the circuit in (b), note that the output signal from each collector of Q_1 and Q_2 is coupled to the base input of the other transistor. Since the common-emitter circuitry in amplifiers provides for a phase reversal of the signal between base and collector, the output of Q_1 is out of phase with the input at its base. This out-of-phase signal is applied to the base of Q_2, and when the signal appears at the collector of Q_2, it is again out of phase with that at the base, thus bringing the phase back to the original phase condition existing at the input of Q_1. Thus by cou-

Vertical Sweep Blocking Oscillator / 151

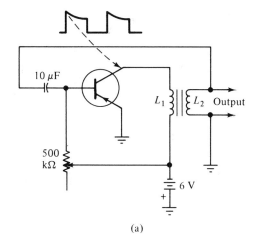

(a)

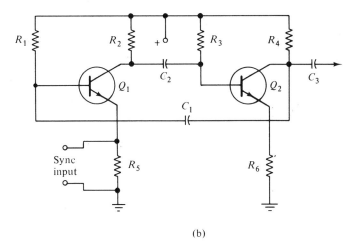

(b)

Figure 5-6 Relaxation oscillators

pling the collector of Q_2 back to the base input of Q_1 (with capacitor C_1), an in-phase feedback loop for producing and sustaining oscillations will be formed.

In the design of a multivibrator such as that in (b), it must be kept in mind that the circuit is symmetrical, having resistors of identical value in each base circuit and collector circuit, as well as capacitors of similar value. Transistors are of the same type with equal characteristics. Despite symmetrical design, however, when power is first applied, current flow rises more rapidly in one transistor circuit than in the

152 / Oscillator Circuits

other, and this slight unbalance is sufficient to initiate the oscillatory cycle.

A synchronizing input signal is applied across resistor R_5 as shown, and when the synchronizing signal is at or near the free-running frequency of the multivibrator, synchronizing lock-in is achieved. If the sync pulse signal arrives approximately between the pulse formation in the multivibrator, synchronization will fail.

5-9 VERTICAL SWEEP BLOCKING OSCILLATOR

Figure 5-7 shows a vertical sweep waveform generator using the blocking-oscillator principle. A sync pulse input is applied across the primary L_1, and two secondary windings, L_2 and L_3, are used for the blocking oscillator circuit. As shown, L_2 in series with the collector terminal supplies the feedback inductance and, coupled to L_3, furnishes the regenerative signal required for oscillations. The sweep output signal is obtained from across R_2 as shown, and this is applied to the vertical sweep output amplifier or, if needed, to an intermediate driver circuit.

Resistor R_1 is the hold control, which is adjusted to maintain sync stability and hence prevent the picture from rolling vertically. This resistor adjusts the forward bias applied to the emitter and, by virtue of the impedance change associated with L_3 and C_1, alters frequency. Resistor R_3, in series with the decoupler resistor R_4, regulates the voltage applied to the emitter, and hence the current through the emitter-collector circuit. Thus, the amplitude of the output waveform is adjusted, and

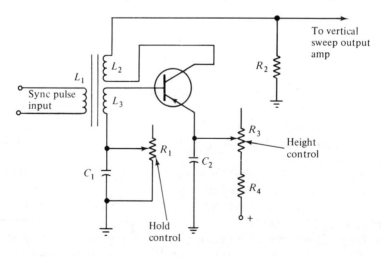

Figure 5-7 Vertical sweep blocking oscillator

thus the height of the pattern appearing on the screen may be regulated. The disadvantage of such a circuit is the necessity for the three-winding transformer; such a transformer can be eliminated by using the multivibrator circuit. The only disadvantage with the latter is the necessity for using an additional transistor, as shown in Fig. 5-6.

In design, the primary L_1 matches the output from the sync-pulse circuitry (usually a sync-pulse separator). Secondary winding L_3 matches the input impedance to the transistor; however, L_2 is not related to the output impedance, but rather to the magnitude of the signal that must be fed back to L_3 inductively to initiate and sustain oscillations. Winding L_2 *must be properly phased* in order for regenerative feedback to be achieved. The amplitude of the magnetic fields produced in L_2 is related to the number of turns and the amount of current flow through the inductor. Also of significance is the degree of coupling between L_2 and L_3. Generally, a minimum of windings should be used for L_2, consistent with initiating and sustaining oscillations.

5-10 MV VERTICAL SYSTEM

Often, in television receivers, a multivibrator (MV) circuit is used for a combination of sweep signal generator and output amplifier. Typical of such a circuit is that shown in Fig. 5-8. Here, transistor Q_3 serves as the vertical sweep output amplifier, with the feedback loop consisting of capacitor C_1 and resistor R_3. This loop originates from the output collector circuit and feeds back to the base input of Q_1 as shown. The vertical driver transistor Q_2 is an emitter-follower circuit, and consequently the signal at the emitter is in phase with that at the base. The output obtained at the emitter is fed to the base of Q_3, and thus the phase is not altered between the output of Q_1 and the input of Q_3. Thus, it is as though the driver were not present and a two-transistor multivibrator circuit were in operation. Since the phase is maintained for proper feedback, the multivibrator generates the required sweep signal, and is kept locked in synchronization by the application of a sync pulse to the base input of Q_1.

Resistor R_6 regulates the bias between the emitter and base, and hence regulates the signal gain of the driver Q_2. Consequently, because the transfer ratio is altered, the height of the sweep is regulated. The linearity control (R_9) is related to the sawtooth modification circuitry and thus changes the sweep-linearity incline. (Sawtooth modification is discussed in Sec. 5-11.)

When a beam in an oscilloscope or a picture tube is made to sweep in a specific direction, it leaves a retrace line in its rapid return to the point of origin. To blank out such retrace lines, portions of the sweep

154 / Oscillator Circuits

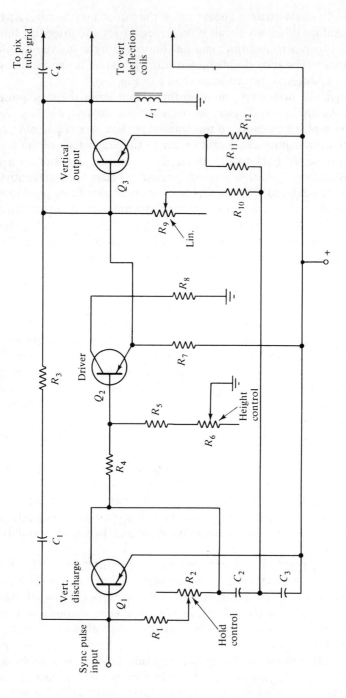

Figure 5-8 Multivibrator vertical sweep system

signals are applied to the grid circuit of the picture tube to blank out the beam during the retrace interval. For the circuit shown in Fig. 5-8, retrace blanking is applied to the picture tube by coupling capacitor C_4.

The output of the vertical amplifier circuit has low impedance and hence is often coupled directly to the vertical deflection coils. A choke coil L_1 completes the collector circuit to ground for the required negative potential for reverse bias.

5-11 SAWTOOTH MODIFICATION

In oscilloscopes the beam is swept across the face of the picture tube electrostatically. That is, the beam sweep is influenced by electrostatic fields created by voltages placed on deflection plates surrounding the beam. Such a system is not feasible for television picture-tube deflection, because the tube length is increased and linearity problems occur when the beam sweeps towards the edge of larger-size picture tubes. Magnetic deflection is employed, but this causes problems because of the necessity for a sawtooth-type sweep in order to draw the beam across the face of a picture tube linearly. Hence, it is necessary to modify the sawtooth signal voltage in order that a linear sawtooth of *current* will be developed in both the vertical and horizontal deflection coils. The linear sawtooth of current is required because a magnetic field that rises and collapses in linear fashion must be produced by the coils.

The voltage-current relationships of magnetic deflection are illustrated in Fig. 5-9. In (a) is shown a high value of resistance, with a small-value inductance in series. Because the inductance value is too low to have an appreciable effect on the relative phase between voltage and current, it produces a sawtooth of current when a sawtooth of voltage is injected into the input terminals as shown.

When, however, inductance is much higher in value in relation to the ohmic value of the resistance, the current in the circuit will lag behind the voltage, and no longer will produce a linear sawtooth of current. The phase difference comes about, of course, because the predominant inductance causes a phase shift between voltage and current. Now, in order to obtain a sawtooth of current, a squarewave must be inserted, as shown in Fig. 5-9(b).

In television the horizontal and vertical deflection coils (called a *yoke* when referred to in combination) have some resistance present in the windings. Such resistance combined with the inductance necessitates the use of an input voltage that combines the sawtooth with the squarewave, as shown in Fig. 5-9(c). This input waveform appears as though the sawtooth had been placed over the top of the squarewave or

156 / Oscillator Circuits

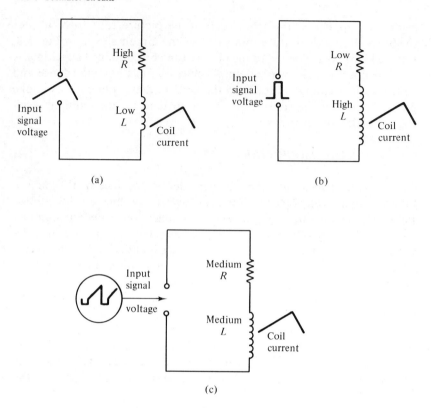

Figure 5-9 Circuitry of sawtooth *E-I* relationships

as if the right-hand side of the squarewave's flat top had been raised below the left-hand side. The waveform is shown at the left in (c).

The degree of modification is difficult to predict mathematically, since it relates to the exact resistance-inductance factor found in the deflection coils. Sufficient modification must be applied so that linear sawtooth is obtained on the screen as indicated by a crosshatch generator, as shown in Fig. 5-10 and described later in this section.

For Fig. 5-8, note that a resistor (R_{11}) taps the emitter of output transistor Q_3 and couples this voltage to the junction of capacitors C_2 and C_3, in the base circuit of the vertical transistor Q_1. Thus, a portion of the waveform in the emitter is fed back to combine with the input signal so as to modify it to proper degree. Consequently, a circuit that modifies and forms the proper sawtooth is often referred to as a *discharge* circuit, since it uses a capacitor for building up a gradual charge (representing a linear rise of the sawtooth voltage) and provides a pulse input for triggering the circuit for a rapid discharge of the sawtooth-forming capacitor.

Horizontal sweep systems in television receivers employ sweep oscillators in a way similar to those used for the vertical systems. For the vertical systems in black and white receivers, the sweep rate is 60 Hz, whereas for horizontal systems it is 15,750 Hz. For color television, the vertical scan frequency is 59.94 Hz, and the horizontal scan frequency is 15,734 Hz. The color sweep frequencies, however, are sufficiently close to those in black and white that no synchronization problems occur when the color transmission is received on a black and white receiver, or when black and white transmission is received on a color receiver. In a properly adjusted receiver, sweep stability should be maintained for either type of reception.

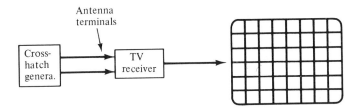

Figure 5-10 Crosshatch pattern for sweep-linearity check

For the illustration in Fig. 5-10, a crosshatch pattern is obtained by utilizing a commercial-type instrument applied to the antenna terminals of the television receiver as shown. Straight-line formations should be present both vertically and horizontally, as shown in the figure. If the vertical lines bend or otherwise distort, vertical linearity is not true. Similarly, if there is bending or crowding of lines horizontally, distortion is present.

CHAPTER 6

Special Circuits

6-1 FLIP-FLOP CIRCUIT

The flip-flop circuit shown in Fig. 6-1 is widely used in electronics because of its ability as a switching device, a gating and counting unit, and a pulse source of either polarity. The circuit is also termed an *Eccles-Jordan circuit* after the originators. One of its important applications is in digital computer circuitry, where it is used for arithmetic operations and number storage, and to provide open- or closed-circuit conditions as required. The circuit resembles the multivibrator relaxation oscillator described earlier in Sec. 5-8. Actually, it is not a generator or oscillator because it does not produce a continuous output, but only develops a signal change at the output when an input pulse is applied.

The circuit shown in Fig. 6-1 is also referred to as a bistable device because it has two stable states; one is the "off" state and the other the "on" state. In the "off" state, the circuit represents zero and a so-called "set" pulse is applied at the terminal shown to trigger the circuit into the "on" or "one" state. A reset pulse applied to the terminal feeding diode D_2 brings the circuit back to zero again. By using binary (base two) arithmetic notation, mathematical functions can be performed by a series of flip-flop stages in cascade.

The flip-flop circuit is symmetrical; hence resistors R_2 and R_5 are identical, as are resistors R_1 and R_6 as well as R_3 and R_4. Similarly, C_1 and C_2 have the same capacitance value. Also, both transistors should be well matched, as should the two diodes in the set and reset lines.

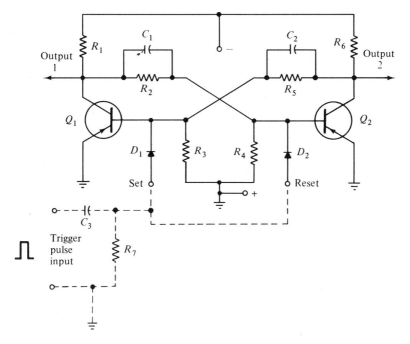

Figure 6-1 Basic flip-flop circuit

If we assume that the 0 state occurs when a reset pulse is applied to the terminal shown, a positive pulse arrives at the base of Q_2 and overcomes any forward bias that exists between emitter and base. Consequently, transistor Q_2 is driven into the nonconduction state.

Note that the power supply potentials are across resistor R_1, in series with R_2 and R_4 to ground. Thus, these three resistors form a shunt resistive circuit across the power supply. Also, resistors R_6, R_5, and R_3 form another shunt network across the power system. Forward bias for Q_1 is obtained from across R_3, which is negative at the base with respect to the positive ground.

When the reset pulse triggers Q_2 into nonconduction, the current through the emitter and resistor R_6 decreases and only small-value currents flow through the rest of the series string composed of R_5 and R_3. Consequently, very little negative potential develops across R_6 and R_5, and virtually the full negative-potential supply voltage is then transferred to the base of Q_1. This supplies Q_1 with the necessary forward bias and causes this transistor to remain in conduction. While transistor Q_1 conducts, high currents flow through R_1, and a large voltage drop occurs across this resistor. Also, the conduction of Q_1 creates a low impedance between its collector and emitter, thus placing a low-resistance shunt across the series resistors R_2 and R_4. Hence, the negative potential at the base of Q_2 is too low to permit conduction, and this transistor remains cut off. The circuit will now remain in this "off" state until it is triggered into its second state.

If a *set* pulse of positive polarity is now applied to the base of Q_1, it overcomes the negative forward bias, thus driving this transistor into cutoff. The voltage drop across R_1 now declines, and Q_1 no longer behaves as a low-impedance shunt across R_2 and R_4, since it is in the nonconducting state and hence acts as an open circuit. Now, Q_2 obtains the necessary forward bias, and conduction occurs again. During conduction Q_2 acts as a low impedance across resistors R_5 and R_3, which, in conjunction with the increased voltage drop across R_6, reduces bias to Q_1 and keeps it in the nonconducting state. Thus, the other state (1) has been reached for the flip-flop, and it will remain in this state until a reset pulse is applied.

When the flip-flop is either set or reset, current changes occur in both transistors. For each trigger input, the current for one transistor rises, and the current for the other drops. Thus, the *current change* through the collector resistor alters the collector voltage, producing an output pulse of one polarity from Q_1, and another pulse of opposite polarity from the collector of Q_2. Depending on the particular application of the flip-flop circuit and the polarity of the signal needed, the output is obtained from one particular collector line and applied to successive flip-flop stages in cascade as required.

The broken-line section shown underneath the *set* and *reset* terminals is for application of successive pulses to perform the set and reset functions without having to switch the pulse from one line to the other. As shown, the diodes permit the entry of only positive pulses to the base elements and also isolate the bases from each other. When a positive triggered pulse is applied to the input (capacitor C_3 and resistor R_7), the pulse appears at both base elements. Since, however, one base is already positive, the incoming pulse has no effect there. For the other base (negative) the forward bias is eliminated by the incoming positive pulse, and the transistor is driven into cutoff. Hence, successive pulses alternately turn the flip-flop to its zero state and then to its one state, back to zero, etc.

Capacitors C_1 and C_2, which shunt R_2 and R_5, permit faster switching rates, since they are in series (to ground) with interelement capacitances and thus effectively decrease total capacitance factors and charging rates. The interelement capacitances introduce delays in trigger and reset speeds because of the time interval involved in the charge and discharge rates of their storage capabilities.

6-2 ONE-SHOT MV

A one-shot multivibrator is illustrated in Fig. 6-2. This is a monostable circuit that, when triggered by an input pulse of proper polarity, produces an output pulse of a predetermined duration. It differs from the astable multivibrator, because it is not free-running or continuously generating a signal. Instead, the output pulse is obtained only when a pulse is applied to the input. Because of the predetermined width of the output pulse, this circuit is useful for converting a train of pulses (with various widths) to a pulse train in which each pulse has the same width as the other. It can also be used as a delay circuit, to delay pulses for a certain time interval.

For the circuit shown in Fig. 6-2, *npn* transistors are used. Note the application of a negative potential to the bottom of R_5. This applies a negative potential to transistor Q_2 that is opposite to the polarity required for the necessary forward bias. The application of such a reverse bias to Q_2 cuts off this transistor, and hence conduction ceases. During this time there is no voltage drop across the collector resistor R_4; hence the potential at the collector of Q_2 is equal to that of the voltage source terminal T_1. Transistor Q_1, however, has the necessary positive forward bias supplied to the base by resistor R_3. The negative potential for the emitter is obtained through the ground circuit, which is common to the

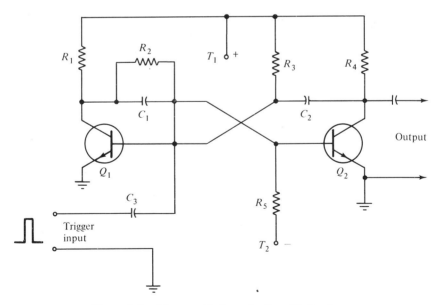

Figure 6-2 Monostable (one-shot) multivibrator

negative terminal of the supply feeding terminal T_2. Thus, transistor Q_1 conducts at saturation, and Q_2 is cut off.

For the circuit shown in Fig. 6-2, a positive-polarity pulse is required for triggering. When this is applied to the trigger input, it feeds (via C_3) a momentary positive polarity directly to the base of Q_2. This establishes the required forward bias for this transistor, and conduction occurs. Now current flows through R_4, and a voltage drop occurs across this resistor. Consequently, the collector potential declines toward the ground level. This voltage change toward the negative potential is coupled by capacitor C_2 to the base of Q_1, and hence reduces the forward bias and decreases conduction for transistor Q_1. Thus, the voltage drop across R_1 decreases and causes the collector voltage for Q_1 to rise toward the positive T_1 polarity. The increase in positive potential applied to the base of transistor Q_2 (via R_2) increases the forward bias on this transistor and increases conduction. As the changes occur, in a brief time interval transistor Q_1 is driven to cutoff, while Q_2 is conducting fully.

Since the trigger pulse initiating the change has only a short duration, the positive potential that has been applied to the base of Q_2 drops to zero as the pulse duration ends. Thus, in a brief interval of time capacitor C_2 assumes its original charge level and permits transistor Q_1 to go into conduction again. When this occurs, the decreased voltage at the collector of transistor Q_1 no longer is capable of overcoming the reverse bias applied to the base by the negative supply potential at terminal T_2, so Q_2 cuts off. The original state prevails again, and the conduction-to-nonconduction period of Q_2 produces an output pulse having a duration controlled by circuit constants.

6-3 RAM AND ROM

There are many types of memory devices utilized in computers, including ferrite cores, tape, magnetic discs, and flip-flop circuits. When circuits such as the flip-flop units are used, information can be read in, stored temporarily, and read out as required. Such memory systems are known as *random-access memories,* and are also known as RAM units. Other memory systems have information implanted in them of a permanent nature that cannot be erased. Thus, a series of circuits such as the monostable multivibrator of Fig. 6-2 could be utilized, for instance, to generate positive signals or negative signals as desired. If a series of five such circuits were used, and only the first, third, and fifth circuits produced an output, the binary number 10101 (21) would be produced. Thus, this set of five could be triggered at any time, and this number would always be available. This is the basis for the *read-only memory*

(ROM). Such memory systems are useful in computers and calculators for performing specific mathematical functions such as square root, trigonometric functions, etc.

Such circuitry is incorporated in the integrated circuits and modules described in Chapter 7.

6-4 INTEGRATOR CIRCUITRY

Signal-modifying circuits are extensively used in all branches of electronics for proper shaping of waveforms. Typical is the integrator circuit shown in Fig. 6-3. Such a circuit can attenuate high-frequency signals or diminish high-frequency signal components of pulse-type waveforms. How much attenuation is achieved depends on the capacitive reactance, and hence on the frequency of the signals and capacitance present. Thus, though circuits of this type are used for signal modification, as described in this section, the characteristics of such circuits should be avoided in general design, so that good signal-transfer characteristics are maintained. High fidelity for signal reproduction is affected when shunt capacitances exist between wires, between solid-state elements, and between other circuit components. Such stray capacitances should be kept to a minimum by careful component selection and wiring layout.

The basic integrating circuit shown in Fig. 6-3 uses a series resistor and a shunt capacitor as shown. Such a circuit can be considered equivalent to a low-pass signal filter. For sinusoidal signals there will be progressive attenuation of signals of higher frequencies, and some phase shifting can occur for sine wave signals making up a composite-signal waveform. For pulse or squarewave signals, the waveform is altered because of the attenuation of the higher frequency harmonic components, which all combine to form the structure of the waveform.

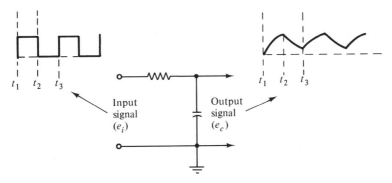

Figure 6-3 Integrator circuit

164 / Special Circuits

From basic electronic theory we know that when the applied voltage across a capacitor is dc, electrons flow on to one capacitor plate and away from the other as the capacitor charges. For ac-type signals such as sine waves or squarewaves, the repetitive polarity reversals cause periodic charges and discharges of the capacitor in opposite directions, the repetitivity conforming to the frequency of the applied signal. Electron flow to and from a capacitor can be considered a capacitor signal current. From the calculus, the applied signal voltage e_c and the capacitor signal current i_c have the following relationship:

$$e_c = \frac{1}{C} \int i_c \, dt \tag{6-1}$$

where e_c = signal voltage across the capacitor
C = capacitance in farads
i_c = capacitor signal current

For this equation, e_c is proportional to the time interval of $1/C$. In practical integrator circuit design the time constant is long compared to the pulse width. Because a pulse contains a high order of harmonic signal components in addition to the fundamental-frequency signal, the harmonic signals may affect calculations. Hence, when the upper harmonic components reduce the capacitive reactance of the capacitor to the degree where the resistance value is much greater, the following equation applies:

$$e_c = \frac{1}{RC} \int e \, dt \tag{6-2}$$

Hence, for the circuit shown in Fig. 6-3, the output signal voltage is proportional to the *integral* of the input signal *current*. The results of integration are shown at the output of the circuit in Fig. 6-3. Positive input pulses are shown, though integration occurs for pulses of either polarity. For the initial pulse at time T_1, the steep leading edge occurs in a short time interval of the applied voltage. The flat top of a pulse maintains the peak voltage across the circuit for a time interval equal to the duration of the pulse. For a capacitor, the voltage buildup lags the current, and voltage rises exponentially. During one time constant ($R \times C$) a capacitor obtains about 63 percent of full charge, with approximately five time constants needed to reach the full charge. However, because the integrator has a long time constant compared to the signal pulse width applied at the input, the capacitor charge voltage does not attain the maximum leveling-off value. Instead, the gradual incline waveform is as shown at the output of the circuit in Fig. 6-3.

When the trailing edge of the pulse arrives, at time T_2, the signal voltage amplitude drops to the zero level in a very short time interval. Now the capacitor discharges through the series resistance and through any input impedance present for the circuit feeding the integrator. The rate of such discharge is again slow compared to the trailing-edge time interval, and the result is an integrated-type waveform as shown. Such a waveform gives evidence of attenuation of higher-frequency signal components, and the actual waveshape depends on the precise time constant used in relation to the width of the input pulse.

The signal input for Fig. 6-3 consists of a pulse train, where the intervals between the pulses have the same width as the pulses. If the intervals between the pulses were longer, it would permit the capacitor to discharge fully between pulses. If the time interval between pulses were much shorter, complete discharge would not occur. For short-duration pulses with shorter time intervals between each, a gradual rise of voltage occurs across the integrator capacitor. This system is often used as a frequency divider, because the firing level of a relaxation oscillator is reached only after a designated number of input pulses have arrived to build up the charge to that required. This circuit is also used in television receivers, for the vertical system, where a number of vertical (short-duration) pulses must arrive before the integrator builds up a sufficient voltage to trigger and synchronize the vertical sweep oscillator.

6-5 DIFFERENTIATOR CIRCUIT

Another signal-modifying circuit has a configuration opposite to that of the integrator, and is known as a differentiating circuit. Thus, as shown in Fig. 6-4, a series capacitor is in the circuit, and the output has a shunt resistor. The series capacitor has an increasing reactance for lower frequency signals, and such a series capacitor attenuates low-frequency signals to the degree of reactance present. Thus, when ampli-

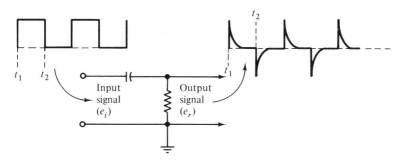

Figure 6-4 Differentiator circuit

fiers using coupling capacitors are designed, higher-value capacitors should be specified for interstage coupling to minimize low-frequency losses. Direct coupling can also be used, without an intervening capacitor or transformer.

When a differentiating circuit is formed, the time constant is made short with respect to the width of the applied pulse. The output is in the form of sharp, narrow pulses with steep leading edges that are ideal for precise triggering and switching purposes in electronic systems. Thus, the differentiating circuit retains the leading edge of the input pulse in the same polarity relationships and, basically, the circuit is equivalent to a high-pass filter, because it attenuates the low-frequency components of pulses or the lower frequency signals of composite signal waveforms.

For the circuit in Fig. 6-4, the applied signal voltage produces a current flow that is proportional to the time derivative of the voltage appearing across the capacitor. The following equation expresses this mathematically:

$$i = C \frac{de_c}{dt} \tag{6-3}$$

Because a differentiating circuit has a short time constant ($R \times C$), some signal components of the input pulse cause the reactance of the capacitor to decrease to a value much lower than the ohmic value of the resistor. (The resistor has a fixed value, but reactance depends on frequency and capacitance.) Thus, the voltage across the resistor in Fig. 6-4 is

$$e_R = iR = RC \frac{de}{dt} \tag{6-4}$$

For the circuit of Fig. 6-4 the leading edge (time T_1) of the input pulse is a quickly rising voltage. In a capacitor, however, current leads voltage, so initially current flow is high for the capacitor. This flow, through the output resistor, produces a sharp leading edge of the differentiated output pulse waveform, as shown for time T_1 at the output. The steady-state input voltage representing the flat top of the pulse provides no change; hence the capacitor current starts to decline as the capacitor charges at a time rate depending on the time constant of the circuit. Because of the short time constant, the capacitor charges rapidly and current flow stops quickly. Consequently, the voltage across the output resistor drops to zero as soon as the capacitor has become fully charged, and current flow ceases.

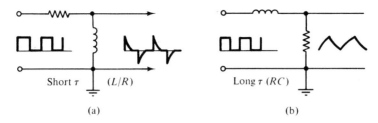

Figure 6-5 *L-R* waveforms

When the trailing edge (sharp voltage decline) of the input pulse arrives at time T_2, the voltage applied to the input drops to zero, and the capacitor now discharges. The discharge direction at the output (time T_2) is opposite to the charge direction. Current flow through the resistor is now opposite to that of the charging current. A negative-spike signal is produced as shown.

Integrators and differentiators can also be formed from resistance and inductance circuits, as shown in Fig. 6-5. The differentiator shown in (a) has a series resistance and a shunt inductance, whereas the integrator in (b) has a series inductance and a shunt resistance. Note that for the differentiator, the resistor is in series with the circuit, as opposed to the type shown in Fig. 6-4, where the resistor is in shunt. Similarly, for the integrator in (b), the resistor is in shunt, whereas for the integrator shown in Fig. 6-3, the resistor is in series. The respective differentiating and integrating output waveforms are similar to those shown for capacitor-resistor circuits.

The RC type of circuits are preferred, because the inherent internal resistance of an inductor disturbs circuit characteristics and makes for difficult design.

6-6 COMBINED *I* AND *D* MODIFIER CIRCUITS

Because the output waveforms from either the integrating or differentiating circuits are related to the time constant, pulses having different widths produce output waveforms that vary accordingly. Thus, these circuits can be used to alter one type of pulse more than another pulse when different types of pulses occur at the input. When, however, sine wave signals are passed through either the integrator or the differentiator, the shape of the sinusoidal waveform is not affected, though there is a change in phase and amplitude.

A single circuit combining the characteristics of the integrating and differentiating circuits is shown in Fig. 6-6. In (a), the output is obtained from across resistor R_2 and is differentiated as shown. In (b), the

168 / Special Circuits

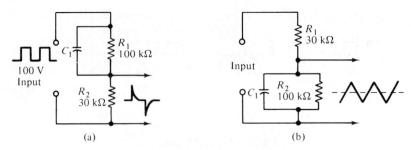

Figure 6-6 Combination circuit

signal is taken from across the parallel RC network and is an integrated triangular waveform as shown. Such a circuit is used on occasion in television or radar receivers for filtering noise pulses while permitting other signals to pass through. It is also useful in cases where long-duration pulses must pass through a circuit while short-duration pulses are kept out.

For the circuit in Fig. 6-6(a), when a pulse first is applied to the input, capacitor C_1 (for a very brief time interval) draws a maximum of current and hence acts as a virtual low-reactance shunt across resistor R_1. Such a momentary shunt prevents any voltage drop across the capacitor. As the latter charges, however, less current flows to it, and the voltage drop across R_1 rises. After five time constants the capacitor is considered fully charged, and the peak pulse potential appears across the *series-resistor network*. Voltage division across the two resistors divides according to Ohm's law; hence for the circuit in Fig. 6-6(a) approximately 66 V appear across R_1 and approximately 33 V across R_2. Hence, the capacitor has a charge across it equal to the voltage drop across R_1. The initial output waveform across R_2 then produces the typical spike characteristic of a differentiated signal.

After the flat-top portion of the input pulse, its amplitude drops to zero, and the capacitor now discharges across both resistors, because they are essentially in parallel with the capacitor, as shown by the equivalent drawing in Fig. 6-7. During the discharge period, reverse current flows through R_2 and produces the negative spike for the com-

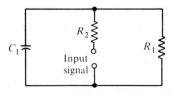

Figure 6-7 Equivalent circuit

pletion of the differentiated output signal. Because the resistors are essentially in parallel with the capacitor, the time constant (τ) for this circuit in equation form resembles the parallel-resistance formula:

$$\tau = \frac{R_1 R_2 C_1}{R_1 + R_2} \tag{6-5}$$

Example: What is the time constant of the circuit shown in Fig. 6-6(a) if the capacitor has a value of 0.02 μF?

Answer:

$$\frac{100,000 \times 30,000 \times 0.02 \times 10^{-6}}{100,000 + 30,000} = 0.00046$$

For the integrator output as shown in Fig. 6-6(b), the resistor R_1 is substantially higher in value than R_2; hence a triangular integrated waveform is produced. The differentiated waveform in (a) is also available with a larger value of R_1 compared to R_2, since most of the pulse voltage appears across the $C_1 R_1$ combination circuit

A circuit that utilizes dual capacitors can also be used, such as that shown in Fig. 6-8. Here, again, the equivalent circuit is the same as that shown in Fig. 6-7, except that resistor R_2 is replaced with capacitor C_2. Hence, the pulse input circuit, in combination with C_2 in series, shunts the parallel circuit consisting of C_1 and R_1. Because, however, capacitors in parallel have a total capacitance equaling the sum of the individual capacitances, the time-constant τ equation for the circuit shown in Fig. 6-8 now is

$$\tau = (C_1 + C_2) R_1 \tag{6-6}$$

Example: For the circuit shown in Fig. 6-8, what is the time constant?

Answer:

$$\tau = (0.0025 + 0.003 \times 10^{-6}) 500,000$$
$$= 0.00275$$

At the entrance of a pulse to the input of Fig. 6-8, the voltage divides across each capacitor inversely to the value of the capacitance for the total pulse voltage. If the pulse duration exceeds five time constants, capacitor C_1 will discharge across R_1, and C_2 will charge to the magni-

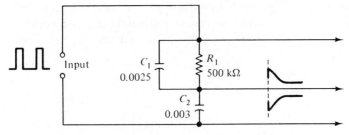

Figure 6-8 Dual capacitor pulse modifier

tude of the pulse voltage. The charge waveforms are as shown in Fig. 6-8 at the output terminals.

Capacitor C_2, at full charge, will discharge across resistor R_1. Capacitor C_1, being in parallel, also has a full voltage of C_2 impressed across it, and both thus discharge through R_1. Because of the parallel-circuit equivalent here, the time constant is, of course, the same for the discharge cycle as for the charge cycle.

6-7 COLOR-KILLER CIRCUITRY

A special circuit used in color-television receivers is the *color killer,* which inhibits signal passage through color circuits during reception of black-and-white picture reception. By blocking signal passage through the color circuitry during this time, picture contamination is prevented.

As shown in Fig. 6-9, a phase-detector circuit is used, as was the case with the phase-locked loop system discussed earlier and illustrated in Fig. 5-4. The circuit of Fig. 6-9 senses the presence of a 3.58-MHz signal input, and since the presence of such a signal indicates color reception, the circuit permits the bandpass amplifier and other color-related circuitry to function normally. In the absence of the incoming burst signal, the detector produces a voltage that alters the bias on the bandpass amplifier transistor, and prevents conduction for this circuit.

The killer phase detector applies zero (or reverse) bias to the base of the color-killer amplifier transistor, preventing conduction during the presence of a burst signal input to the phase detector. A series-resonant circuit for 3.58 MHz (composed of L_3 and C_4) feeds a signal of such a frequency from the 3.58-MHz oscillator or amplifier to the junction of the phase detector diode. Thus, the phase detector compares the frequency and phase of the 3.58-MHz signal with that of the incoming burst signal of the same frequency. While these two signals are present, no voltage change occurs for the bias setting to the color-killer ampli-

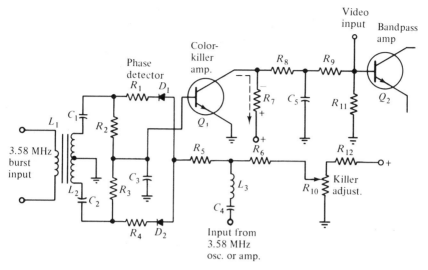

Figure 6-9 Color killer

fier. The bias characteristic is set by the killer adjustment potentiometer R_{10}, which regulates the balance and hence the phase relationships of the detector system.

The forward bias for the bandpass amplifier *npn* transistor is obtained via resistor R_7, and makes the base of Q_2 positive with respect to its emitter. Hence, the bandpass amplifier operates in normal fashion and processes the video signal applied to the base input circuit. During the reception of a black-and-white picture, no burst input signal arrives at the killer phase detector; hence an unbalanced condition prevails, and a positive voltage is applied to the color-killer amplifier transistor Q_1, thus furnishing forward bias. Upon the application of forward bias to Q_1, electron flow now occurs through this transistor in the direction shown by the arrow, and a consequent result is a voltage drop across R_7 because of this flow. The polarity of the voltage drop across R_7 is minus toward the collector side of Q_1 and hence toward the base side of Q_2. Consequently, the reduced forward bias to the base of Q_2 stops conduction of this transistor, and the video signals no longer are amplified by the bandpass Q_2 transistor.

Bias conditions for Q_1 and Q_2 are somewhat critical, and the proper values must be supplied so that the color killer remains in a nonconducting state while color reception prevails, but goes into conduction to shut off the bandpass amplifier upon reception of black-and-white pictures. For the phase detector, symmetrical circuitry design must prevail as with other phase detector systems. Capacitors C_1 and C_2 should be well matched, as should resistors R_1 and R_4, as well as resistors R_2 and R_3. Also, diodes D_1 and D_2 should have similar characteristics.

6-8 AM DETECTION

A typical detector (demodulator) for amplitude-modulated signals is shown in Fig. 6-10. The IF signal at the output of the last IF amplifier is fed to the usual resonant circuit composed of C_2 and L_1. The transformer arrangement between L_1 and L_2 transfers this signal to the detector resonant circuit composed of L_2 and C_3. Diode D_1 serves as a rectifier, which produces single-polarity pulses having an amplitude average corresponding to the audio signal present. Capacitor C_4 filters the pulse-type signal to produce the audio component. Resistor R_3 is a

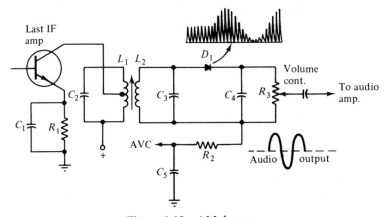

Figure 6-10   AM detector

volume control, which transfers the audio signals to appropriate amplifiers.

The detector system loads down the last IF amplifier stage to a greater degree than would exist if this were followed by successive IF amplifiers. Consequently, the last IF amplifier stage does not have the gain found in the earlier stages. Hence, while the core tuning between L_1 and L_2 is broadly resonant, it should nevertheless be adjusted carefully for maximum output to increase the signal-to-noise ratio.

The lower section of the AM detector is an automatic volume control circuitry, as discussed in the next section.

6-9 AVC

Automatic volume control (AVC) is widely used in radio receivers to compensate for the difference in carrier strength of incoming signals and thus to level them all off to the general amplitude set by the volume control. This is done by sampling the amplitude of the incoming carrier, and obtaining a signal (dc) of a specific polarity for adjustment of the bias of previous stages. Thus, the AVC voltage obtained is supplied to the input circuitry of RF amplifiers, as shown earlier in Figs. 4-7, 4-8, and 4-9.

The AVC takeoff shown in Fig. 6-10 consists of resistor R_2 and capacitor C_5. Resistor R_2 tends to isolate the detector system from the earlier RF amplifier stages; capacitor C_5 has a low reactance for audio signal frequencies, and hence shunts any of these appearing at the AVC output from R_2. Thus, only the average dc representative of the carrier is fed into the AVC line and is applied to the input circuits of the earlier-stage amplifiers.

If a bias of opposite polarity is required, the diode detector D_1 is reversed. If this is a video detector (which is also an AM detection process), the diode could not be reversed for a change of polarity for the AVC, because it would upset the polarity relationships of the signal information reaching the picture tube. (Signals arriving at the picture tube drive the grid into a greater negative-bias region for darkening the picture tube. For incorrect polarity, the opposite would occur, and the picture scenes would become brighter instead of darker.)

6-10 FM DETECTION

A typical frequency-modulation (FM) detector is shown in Fig. 6-11. Here, two diodes (D_1 and D_2) are used in a detector circuit termed a *ratio detector,* wherein inductor L_4 samples some of the signal at the resonant circuit L_1 of the output IF amplifier. Phase comparisons are made in this circuit between the signal picked up by L_4 and that appearing across L_2 and L_3. As the carrier frequency shifts during frequency modulation, the frequency changes are converted to representative audio voltages and obtained from across R_3 (which also serves as the volume control).

During detection, one diode conducts more than the other, depending on the phase relations set up at the input circuit. Thus, the ratio of voltages across the output resistors R_2 and R_3 change, though the total voltage across these two resistors remains the same. Capacitor C_6 shunts the two resistors and during design is chosen to have a high capacitance value (usually several microfarads). Capacitor C_6 charges to the dc potential appearing across the output resistors. Since capacitor C_6 opposes a voltage change, it tends to maintain a constant voltage across the resistor combination. Thus, sudden changes, such as those

174 / Special Circuits

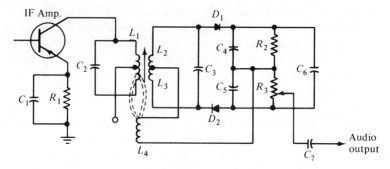

Figure 6-11 FM detector

caused by sharp static bursts or noise pulses, are minimized and often eliminated. A similar circuit, called a *discriminator,* has the two diodes wired in the same direction. The discriminator has been used on occasion in older receivers, but since it requires an amplitude-clipping stage (limiter) for elimination of amplitude modulation in the form of noise, it has been generally replaced by the FM detector shown in Fig. 6-11.

6-11 TV H-V SYSTEM

The high voltage necessary for the second anode of the picture tube is obtained from a system that is an integral part of a horizontal-sweep output system of a television receiver. Typical is the circuitry shown in Fig. 6-12. Here, the signals obtained from the horizontal sweep oscillator and amplifier are applied to the grid of the output transistor, and the amplified version develops across the primary of the horizontal output transformer L_1. Since, however, impedance of the output transistor is sufficiently low to match that of the deflection coils, the collector of the horizontal output transistor is coupled directly to the horizontal sweep coils as shown. (With tube-type receivers, coupling to the coils is usually from a secondary winding.)

In a deflection system such as that shown in Fig. 6-12, high-potential pulse signals are present, as well as the modified sawtooth necessary for a linear current sweep through the deflection coils. Transient voltages result when the fields of the horizontal deflection coils collapse during the sawtooth retrace. To dampen such transients, a damper diode D_1 is used, which is polarized to bypass the transient oscillations only. [The resonant circuit, formed by the horizontal deflection coils plus the distributed capacity of such coils, contributes to the tendency to be pulsed into spurious (unwanted) oscillations.]

The voltage developed across L_1 is stepped up by L_3, and a series

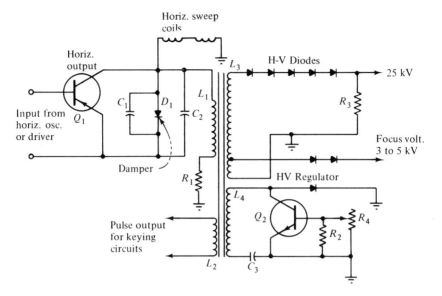

Figure 6-12 TV H-V system

string of silicon rectifier diodes is used so that their combined voltage rating is in excess of the voltage produced for the picture-tube anode. Special precautions must be observed here to minimize arcing, and each diode should have similar characteristics and similar voltage ratings. On some occasions, design engineers bridge the diodes with capacitors to equalize impedances and minimize voltage surges across the individual units.

A tap is provided for the focus-grid voltage as shown in Fig. 6-12. In some receivers the rectifier high-voltage stack is tapped to furnish the required focus-electrode voltage without the necessity for the additional diodes shown in the focus line. Resistor R_3 in the high-voltage line, usually in the neighborhood of 300 megohms, acts as a bleeder resistor for the high voltage. Thus, it provides a slow discharge path for the picture tube anode-surface capacitance, minimizing shock hazards.

Transformer winding L_2 supplies a pulse source for circuits such as the bandpass amplifier, color killer, keyed AGC, and other similar circuits. Winding L_4 is for forming a high-voltage regulating circuit with transistor Q_2. Because of induction between L_3, L_4, and the primary, voltage changes are reflected across L_4 and alter conduction of Q_2. Loading effects are thus utilized by the high-voltage regulating transistor for compensating for the variations that may occur in the high-voltage system. Resistor R_4 shunts the base-emitter terminals, and hence adjusts bias relationships and conduction for Q_2. Consequently,

6-12 DARLINGTON TRANSISTOR

The Darlington transistor symbol is shown in Fig. 6-13(a). It consists of two transistors, the first in an emitter-follower arrangement that feeds the base of the second transistor section. The feature of this unit is its exceptionally high gain. The total signal current amplification (H_{fe}) is the product of the amplification available from each of the transistors within the Darlington structure. Thus, $H_{fe_1} \times H_{fe_2} = H_{fe_d}$. Thus, if the input transistor of the Darlington pair has a beta of 100, and the second transistor a beta of 60, then the total beta would be 6000. In this instance, the 6000 may be slightly higher because of the common collector currents for the driver and the output transistor as shown.

A typical circuit showing associated components is shown in Fig. 6-13(b). Here, resistors R_1 and R_2 form a voltage divider for supplying bias to the base of the first transistor. Resistor R_L, in series with the emitter of the Darlington, forms the output. Such a device has many applications, particularly where high amplification (beta) is required. Sensitivity is high and the output from the Darlington has a collector current sufficiently high to operate as an efficient driver, even though low supply voltages are utilized. The Darlingtons are useful for reducing the number of components which would otherwise be needed in general transistor circuitry.

The Darlingtons are useful in audio circuits, oscillators, and switching devices. There is also an impedance step down between the

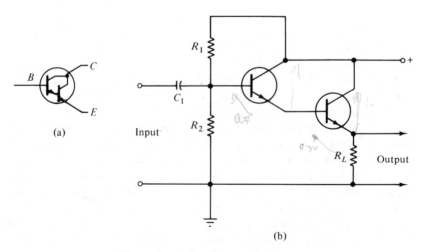

Figure 6-13 Darlington circuitry

input and output, and hence the Darlington simulates the characteristics of a step-down transformer, but with extremely high gain. The approximate input impedance is equal to $\beta^2 \times R_L$. The output impedance is usually equal to the resistance of R_L. In switching applications, Darlingtons are useful to as high as 25 kHz.

6-13 DIFFERENTIAL AMPLIFIER

The basic circuit of a differential amplifier is illustrated in Fig. 6-14(a). As shown, it consists of two transistors having a common-emit-

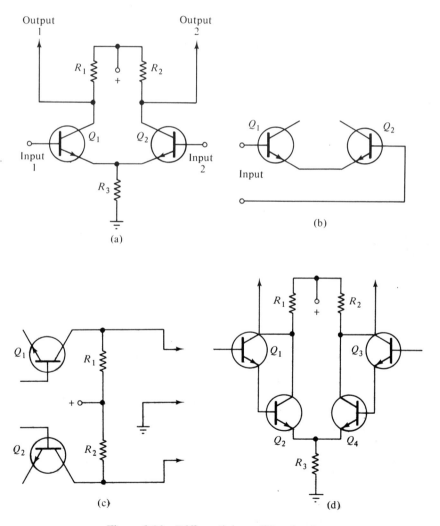

Figure 6-14 Differential amplifier circuits

ter circuitry and sharing resistor R_3. The two transistors are emitter-coupled directly without intervening capacitors. The direct coupling permits a wide-signal bandpass for amplification as well as excellent circuit stability. The differential amplifier is versatile and can serve as a signal mixer, a limiter (by being overdriven by the input signals), a modulator, and also a signal-frequency multiplier. Because of the minimum number of component parts (and the absence of inductors and capacitors), it lends itself well in the design of integrated circuitry of the type described in Chapter 7. It is frequently used in conjunction with the operational amplifier described in Sec. 6-14.

Several modes of operation are possible with the differential amplifier. In one method the signal is applied to the base of only one transistor (with the second input at signal ground). If the input is applied to transistor Q_1 in Fig. 6-14(a), for instance, the amplified version of the input signal appears at the collector of Q_1 with the usual 180-degree phase difference between input and output signals obtained for the common-emitter circuitry. The signal variation also appears across the common-emitter resistor R_3. Since current flow through R_3 is also through transistor Q_2, emitter-collector current changes occur for Q_2 representative of the signal components.

Assume, for instance, that a positive alternation of the input signal appears at the base input of Q_1. This signal then furnishes an equivalent forward bias and causes an increase of current flow in the collector emitter of Q_1. The voltage drop across the collector resistor R_1 increases with the transistor junction at the collector becoming more negative. This is then representative of a negative change, or 180 degrees difference between signals at the input and output for Q_1.

The rise of current through R_3 causes the emitter side of this resistor to become more positive and in turn causes the emitter of Q_2 to have a reverse bias applied. Hence, current through Q_2 declines, and the voltage drop across R_2 decreases. Now the collector of Q_2 becomes more positive, producing a signal at the collector of Q_2 that is out of phase with that of the Q_1 collector. Thus, a phase-splitting characteristic is obtained for this circuit. If the output is taken from the collector of Q_1, the mode is a single-ended input-output signal-inverting mode. If the output is obtained from the collector of Q_2, we have a single-ended input-output noninverting mode.

Instead of applying the input signal to one base terminal only, one can apply the signal *across* the two inputs as shown in Fig. 6-14(b), thus forming an input referred to as a *differential type*. The output can again be obtained from either collector, or from both as against ground for a balanced output as in (c). The circuit in (d) is formed by using two Darlington pairs to form a differential amplifier. Thus the advantages of the

Darlington circuitry are combined with those of the differential amplifier. (See Sec. 6-12.)

A very useful characteristic of the differential amplifier is its *common-mode* characteristic obtained for in-phase signals appearing at both inputs. Such signals at the input are those undesired types that may be picked up by power-supply feed lines, or by spurious signals entering the circuit by coupling or radiation. Since such signals affect both base inputs at the same time and with the same phase, a voltage differential occurs across the R_3 emitter resistor and introduces emitter feedback (current type), which causes attenuation of the common-mode signal, but without affecting a single input signal that is applied for amplification. Because of the common-mode characteristic, the differential amplifier is highly stable. Any voltage fluctuations are ac-type signals and, when they appear at both base inputs, the result is attenuation which minimizes the appearance of such signals at the output.

For maximum performance, the two transistors of the differential amplifier should be well matched in their functional characteristics, and the two collector resistors should have identical ohmic values. Greatest stability and efficiency for the differential amplifier are obtained when the value of the common emitter resistor is increased, since it simulates a high-impedance constant-current source, and effectively reduces interaction between the input and output circuitry of the two transistors. Increasing R_3, however, produces a larger voltage drop and a consequent requirement for the power source to furnish higher collector potentials. A better procedure is to employ the constant-current sources described in Sec. 6-15.

6-14 OPERATIONAL AMPLIFIER

The operational amplifier is a unit especially designed for exceptionally high gain and a flat response with minimum drift. Because direct coupling is used between stages, the amplifier has linear characteristics. Because of these the signal-frequency response is from dc to very high signal frequencies. The feedback control is usually employed with the operational amplifier for modification of the input impedance or the feedback impedance as required. The basic circuit is shown in Fig. 6-15. The feedback loop is designated by the lowercase Greek letter beta (β), and represents the *decimal equivalent* of the *percentage* of voltage fed back. [This symbol is also used for gain, as in Eq. (2-1).]

The feedback, when inverse, is used for the reduction of harmonic distortion and for increased bandpass and noise reduction. (See Section 3-9.) With inverse feedback, signal amplification is reduced in proportion to the amount of the inverse feedback utilized. When a feed-

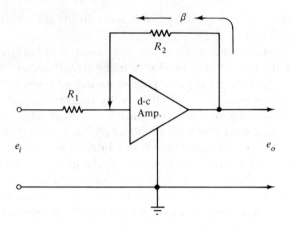

Figure 6-15   Operational amplifier

back signal is applied to the input of an amplifier, it appears at the output of the stage to which it is applied as an amplified signal, but out of phase with the distortion that develops within that stage. In consequence, cancellation of distortion (or noise signals) is obtained to the degree established by the level of the feedback voltage.

Without feedback the input-signal voltage e_i is increased by the amplification A, and the result is an output signal e_o, or, in equation form

$$e_i = A e_o \tag{6-7}$$

Hence, the amplification *without* feedback (also termed the *open-loop* amplification) is the ratio of the instantaneous output-signal voltage and the input-signal voltage:

$$A = \frac{e_o}{e_i} \tag{6-8}$$

In feedback systems the lowercase Greek letter *beta* (β) symbol is preceded by a minus sign when the feedback is inverse or negative. With oscillators, where positive feedback is used for regeneration, the symbol is preceded by a plus sign. Another symbol, A', designates the *signal voltage amplification with feedback.*

The product $A\beta$ indicates the *feedback factor*. Thus, $1 - A\beta$ is a measure of the feedback amplitude. In equation form, the signal-voltage amplification with feedback is

$$A' = \frac{A}{1 - A\beta} \quad \text{or} \quad \frac{A}{1 + A\beta} \tag{6-9}$$

where A' = the signal-voltage amplification *with* feedback
A = the signal-voltage amplification *without* feedback
β = the decimal equivalent of percentage of output-signal voltage fed back.

When the feedback factor $-A\beta$ is much greater than 1, the signal-voltage gain is independent of A, and the equation for signal-voltage amplification with feedback becomes

$$A' = -\frac{1}{\beta} \tag{6-10}$$

Because inverse feedback reduces distortion, it is also convenient to calculate the amount of distortion that is present with feedback. D' indicates the distortion of the output-signal voltage with feedback, and D shows the distortion of the output-signal without feedback. Thus, the equation is

$$D' = \frac{D}{1 - A\beta} \tag{6-11}$$

Thus, both the gain and distortion are reduced to the degree established by the amplitude of the feedback $1 - A\beta$. If, for instance, the latter were 3 and the gain without feedback = 60, the gain would be reduced to 20:

$$A' = \tfrac{60}{3} = 20$$

Had the distortion been 6 percent before the feedback was applied to the circuit, the distortion would be reduced to 2 percent:

$$D' = \tfrac{6}{3} = 2$$

With the feedback factor greater than 1 (where signal voltage gain is independent of the A factor) the output-signal voltage e_o is now influenced only by signal currents in R_1 and R_2 of Fig. 6-15 plus the input voltage e_i. Thus, in a high-gain operational amplifier with feedback, the output-signal voltage is indicated by the following equation:

$$e_o = \frac{R_2}{R_1} \times e_i \tag{6-12}$$

6-15 CONSTANT-CURRENT SOURCE

Figure 6-16(a) shows a differential amplifier similar to those shown earlier in Fig. 6-14, except for the special symbol in the common-emitter line. Here, the intertwined circles represent a *constant-current source*. When such circles are included in a circuit they emphasize the need for taking special precautions regarding current stability to obtain optimum performance. As an alternate to the high-resistance constant-current source (see Sec. 6-13), an additional transistor can be used, as shown in Fig. 6-16(b). Here the impedance between collector

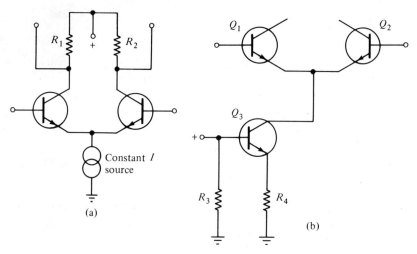

Figure 6-16 Constant-current circuitry

and emitter of Q_3 is sufficiently high for constant-current purposes, yet resistor R_4 can now be a low value to reduce the voltage drop across it and thus to reduce the power dissipation. Also, the demands on the power supply for raising potentials to meet the requirements of the collector circuits have been diminished. The system shown in Fig. 6-16(b) lends itself to IC fabrication.

A modification of the circuit consists of replacing R_3 with a voltage-reference diode that has the ability to compensate for bias changes in Q_3 caused by temperature effects.

CHAPTER 7

Integrated Circuits (ICs) and Modules

7-1 IC BASICS

An integrated circuit (IC) is a microminiature section containing a number of electronic components (such as transistors, resistors, diodes, etc.) properly interconnected and designed to perform a designated electronic function. The IC is formed by microprocessing the components so that they become integrated into a so-called semiconductor monolithic *chip*. When such chips are interconnected to form a unified electronic system (such as an audio amplifier, or digital signal-processing circuits, etc.), the IC is termed a *hybrid* type.

Once the integrated circuits have been designed and go into production, the manufacturing is simplified, because they lend themselves to batch processing, permitting the fabrication of a variety of units at one time. The IC chip is the ultimate in electronic microminiaturization and permits the processing and modification of signals at extremely low signal amplitudes, thus saving enormous amounts of space in the final system. During the initial phases of electronic signal processing in various devices, the signal formation, combinations, and other modifications that are necessary can be done at these low voltage and current levels. When the process is complete, the final signal can be processed additionally to raise it to such levels as are necessary for power amplitudes needed to actuate transducers, power amplifiers, LED readout devices, etc.

The IC units often form a part of a *module* consisting of a signal processor such as a complete audio amplifier, or a complete IF section in a television receiver, etc. Another use for IC units and modules is in the hand-held calculators. As with most IC modules that process complex formations and combinations, the design and manufacturing process is involved. Engineers must design and plan each circuit, as well as the linkage between circuits. Large-scale schematic layouts are usually used; all the components, such as the transistors, diodes, resistors, and capacitors, are accurately positioned in the schematic to depict all electronic functions related to the system under design. The final schematics are reduced photographically to microminiature size on special plastic film, which is deposited on a silicon wafer termed a *substrate*. Various processes are utilized, some including acid etching of the required pattern on the chips, thermal processing, mask fabrication, and chemical etching, in addition to the photo-etching process, diffusion, and vacuum deposition.

Elements and chemical impurities are deposited during thermal processes for forming the semiconductors; the basic principles are covered in later sections of this chapter. Finally, integrated circuitry contains all the components and linkages necessary for performing the intricate mathematical functions required in hand-held calculators, signal generation and timing in digital clocks, etc. Once the IC final product is complete, all that is necessary is that it be linked to the required keyboard assemblies, readout displays, and other final sections of the complete system. The letters LSI indicate "large-scale integration" and refer to such multicircuit ICs.

7-2 IC COMPONENT FORMATION

Virtually all electronic components can be incorporated into the integrated circuit, though not all can be processed at the same level of difficulty. Solid-state devices, such as transistors and diodes, are readily structured within the IC, as are resistors. Capacitors, however, offer some difficulty, and, though their formation is perfectly feasible, the process runs production costs up. Inductors are not possible to produce in the integrated circuit, though *inductance* can be simulated.

During the manufacturing process, the components are interconnected on the same plane or surface of the semiconducting material; this is called a *planar* process. Primarily, it consists of diffusion, oxidation, and the elimination of specific unwanted elements from the final IC chip.

Diffusion, as the name implies, is the intermixing of atomic structures to the degree required so that the combination produces the necessary structure of the specific IC being manufactured. The oxidation process combines oxygen with an element to change the structure.

One basic process is illustrated in Fig. 7-1. In (a) is shown an *n*-type

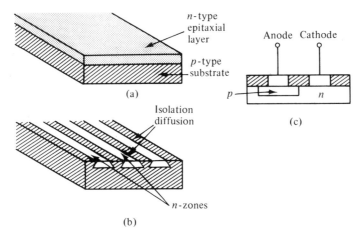

Figure 7-1 Diffusion factors

layer fused with a *p* substrate. The fusion of the two results in an *n* on *p* substrate composite as shown. A *p*-zone impurity is then diffused through the *n* epitaxial layer, as shown in (b). This process forms *n*-type sections, as shown in (b), and the diffusion process is called *isolation diffusion,* because the process isolates the *n*-zone regions by a formed *p-n* junction. As shown in (c), a *p*-type diffusion forms the basic diode, with the anode and cathode terminals connected as shown.

After the sections are formed as shown in Fig. 7-1, a transistor can be manufactured by utilizing *p*-type diffusion to form base regions of *npn* transistors. (The *npn* is most widely used in IC, and it is less difficult to process than a *pnp*. The latter, however, is also found in certain ICs with special applications.)

The formation of an *npn* transistor is as shown in Fig. 7-2(a) and is similar to the formation of a junction diode, except that a third diffusion is undertaken to form the three elements of base, collector, and emitter as shown. The symbol is in (b).

Similar processes are undertaken for the formation of the field-effect transistor. Basically, in the manufacture of the FET unit, a substrate foundation slab of the *n*-type silicon slice is used, with *p*-type source and drain regions diffused into the slab by the use of masking devices. A thin layer of oxide is formed over the surface, and so-called *windows* are opened through an etching process directly over the *p* regions shown in Fig. 7-2(c). The next process is metallic evaporation over the surface. After removal by evaporation, metallic contacts with the source and drain windows are left, plus a metallic span over the peak regions, separated by an oxide coating. With the junction FET, the gate is made up of a *p-n* junction diffused into the channel material.

As shown in (c) the polarities of the slab and regions determine

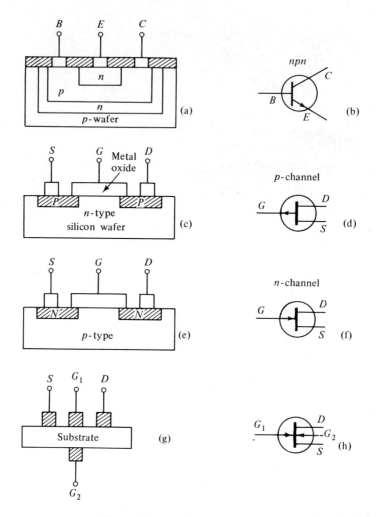

Figure 7-2 Transistor formation

whether the field-effect transistor is a *p*-channel or an *n*-channel type. The *p*-channel shown in (c) has the symbol shown in (d), while the *n*-channel type is illustrated in (e) and (f). (See also Fig. 2-2 and the related discussion.)

A double-gate FET is formed by adding an additional gate element to the slab (but isolated from the first gate). This is shown in Fig. 7-2(g), with the symbol in (h). Again, with a reversal of the diffusion zone polarities, the *p*-channel can be formed as well as the *n*-channel.

Similarly, conductor regions have a specific resistance, and this can be varied by thickness and by the degree of impurities utilized in the manufacture. Resistors having various ohmic values can be fabricated, though values substantially above 40 kΩ produce processing problems and increase manufacturing costs. Ranges below 40 kΩ are most feasible.

An IC containing a diode and a resistor is shown in Fig. 7-3(a). Here, an *n* zone is diffused into a *p* wafer for the diode. Next, a *p*-zone diffusion is formed into the *n* zone.

Reverse-bias *p-n* junctions (diodes) can function as capacitors in

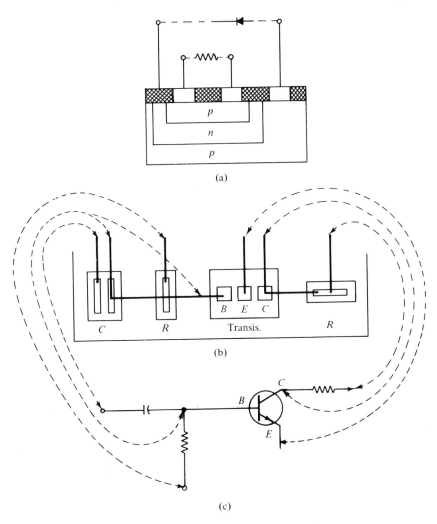

Figure 7-3 ICs and equivalent circuit

integrated circuits because of their inherent interelement capacitances. Such a simulated capacitance, however, has a value dependent on the applied voltage, because of the latter's effect on the depletion region width. This factor is considered during design to provide the required capacitance associated with the proper operation of the integrated circuit. Practical and feasible values range from about 20 pF to about 100 pF.

A section of integrated circuitry is shown in Fig. 7-3(b) and contains two resistors, a capacitor, and a transistor, with interconnections to form a portion of an amplifier circuit as shown in (c). (This illustration is, of course, a highly magnified version of an actual IC.)

An individual integrated circuit formed within the plug-in units may contain hundreds of transistors, diodes, and other components, and yet may have a microminiature size that needs only to contain sufficient bulk to accommodate the output terminals. (General IC packaging is discussed in the next section.)

7-3 BASIC IC TYPES

Linear integrated circuitry of the bipolar types forms amplifying circuits and regulating circuits, and is widely used in audio systems, television receivers, and communications circuitry. Such ICs contain most of the circuit components, though additional resistors, transformers, or other components must usually be added externally to complete the operational system.

Small-scale integration (SSI) contains circuitry such as logic gates in computer systems, or switching devices, using either the bipolar or unipolar construction. Medium-scale integration (MSI) forms counter systems, decoders, mathematical registers in calculators and computers, and similar sections. The large-scale integration (LSI) unifies several sections to form a complete system for performing specific functions. The essential factors concerning such ICs are discussed in subsequent sections of this chapter.

7-4 IC FORMS AND DIAGRAMS

Integrated circuits are packaged in many shapes and forms, some square and others circular, while still others are rectangular in shape. The package units have a number of prongs designated to fit into sockets or into holes in the printed-circuit board for soldering to the metallic conducting strips.

The circular IC package is shown in Fig. 7-4(a). It has a base extension (or notch) that identifies the start of the numbering sequence. For

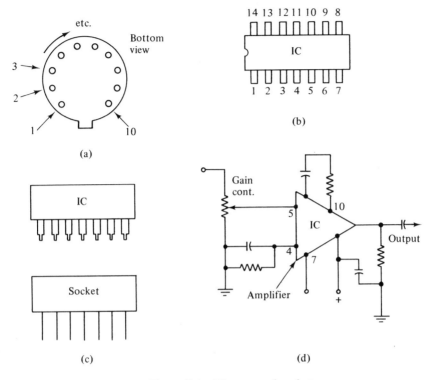

Figure 7-4 IC types and sockets

the bottom view shown, the sequence is clockwise. For the rectangular IC shown in (b), the connecting segments are numbered also in consecutive order, with a notch identifying the start. On occasion, an imprinted dot may be present to indicate the beginning of the count sequence. In (c), the side view of the rectangular IC is indicated, plus the socket often used for convenience in replacement. Such a socket increases the bulk of the installed assembly, but it eliminates the necessity for unsoldering the prongs when the IC becomes defective.

Some of these units measure less than a half inch in length, the actual size depending on the number of circuits within them and the necessity for providing sufficient output terminals with good contacts for plug-in or soldering purposes.

Integrated circuit amplifier units are usually illustrated schematically in triangular form, as shown in Fig. 7-4(d). On occasion, square and rectangular representations are used by certain manufacturers. Only those terminals actively used are numbered in schematics. If more than one IC is used, the usual procedure is to identify them as IC_1, IC_2, and IC_3.

Often several ICs are used to make up a single module, and such a module in a particular system may act as an IF amplifier combined with a detector system and output amplifier circuitry, as shown in Fig. 7-5. Many modules are installed by pressing the units into place, because appropriate prongs and receptacles for such prongs are present for convenience in removal and replacement.

The system shown in Fig. 7-5 is typical of either the sound section of an FM receiver or the FM sound section of a television receiver. The IC_1 contains several stages of IF, and the output is applied across the FM-detector input transformer. This transformer has variable tuning cores for frequency alignment. This section feeds into IC_2, which contains the FM ratio detector, plus one or two audio amplifier stages necessary to raise the signal level to that needed for application to a loudspeaker.

The ICs and the associated circuitry shown in Fig. 7-5 are typical representations of the use of external components of resistance and capacitance for expediency in the design of the integrated circuitry. Thus, where a single resistor routes the power supply potential to several circuits of an IC, the required higher wattage rating may exceed that which it is possible to design within a given integrated circuit. Hence, it can be designed as a part of the external circuitry.

Similarly, it may be preferable in some instances to use external resistors or capacitors to permit some variations either in applications

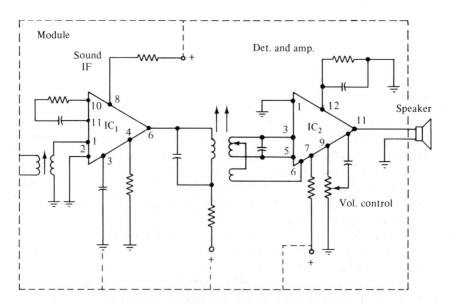

Figure 7-5 Typical ICs and associated circuitry

where the ICs may be biased at different levels for various systems, or by users who need to deviate from basic circuitry. Also, coils present some problems in the simulation of their inductance and hence are used externally. This is also more convenient, because it provides for manual tuning adjustments in RF circuitry.

A single IC may contain several independent circuits, as shown in Fig. 7-6. Here, two audio amplifier sections are present for stereo purposes. The standard triangle symbol for an amplifier is shown for each, plus the terminal connections. (Digital circuitry uses different symbols, as illustrated later in this chapter.)

Typical components and interconnections for the IC shown in Fig. 7-6 are shown in Fig. 7-7. Here, terminal 1 connects to terminal 6 for feedback purposes, using a 1-MΩ resistor. A similar circuit prevails for the other channel, with the feedback from terminal 1 to terminal 9; again a 1-MΩ resistor is used. The input lines and output lines have bypass networks, and 200-μF capacitors are used to couple to the respective speakers as shown. Separate volume controls are used for the input, and 0.05-μF capacitors are in series with the input lines. Small IC units such as this, with the voltage amplitude applied as shown, are capable of delivering approximately 3 W of sine wave signal power to each speaker.

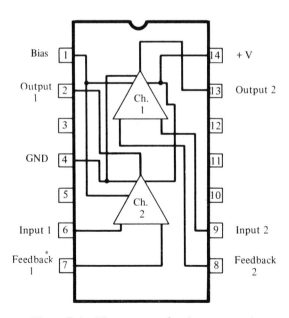

Figure 7-6 IC representation (stereo amps)

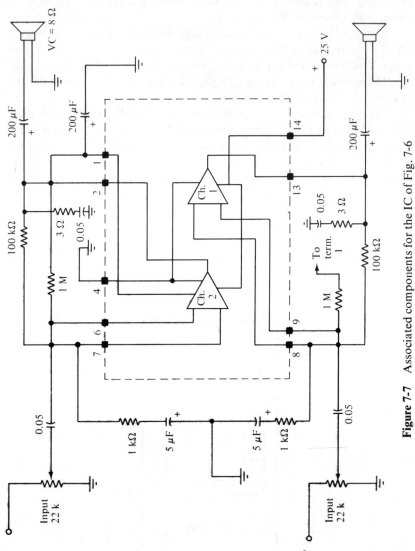

Figure 7-7 Associated components for the IC of Fig. 7-6

7-5 DIGITAL IC LOGIC

In telephone systems, in computers, and in calculators, as well as in communication electronics, a number of logic switching and gating circuits are used. Included with such circuits is the flip-flop circuit already discussed in Chapter 6. Such gating and switching circuits perform logical functions and act as linkages between the various calculating, storage, and other such sections found in computers, microprocessors, and similar systems. Actually, there are less than a half-dozen different *types* of logic switches and gates, but these are used repeatedly to form the bulk of the many that make up a computer or telephone system.

Digital computer systems operate with what is known as *two-level logic;* that is, only two states are utilized. Thus, the circuit conditions of *on* and *off* are utilized to represent one and zero, or, as described in Boolean algebra, the 1 and 0 states are designated as *true* and *false* or related to *yes* and *no* conditions. From these states, certain logical statements are evolved for the switching and gating systems.

For the engineer or technician unfamiliar with computer and switching terminology, the special logical circuits that are employed have unusual names, such as *and* gates, *or* gates, and *not* circuits. The *and* and *or* circuits, in conjunction with the *not* types, perform a variety of useful tasks in routing numbers, switching them as required, and processing them otherwise.

7-6 *OR* CIRCUITS

A typical *or* circuit using diodes is shown in Fig. 7-8(a). As shown, this circuit has three inputs, though more could be used, or (at a minimum) two can be employed as needed. In this circuit no supply potential is required; hence, in the absence of an input signal there is no voltage drop across the output resistor. If a positive voltage or pulse is applied to the A input as shown, in relation to ground, the upper diode conducts, because it is biased in the forward direction. Consequently, current flows through the diode and resistor, and an output voltage develops across the resistor. Hence, an output pulse is produced for an input at A. Similarly, if a pulse is applied at B, there will be an output; the same is true at C. If signal voltages are applied to two or three inputs at one time, conduction, of course, occurs, and an output signal is produced. Thus, an output signal is provided for an input signal at A, or B, or C, or all. (If a negative signal represents logic 1, the circuit has the diodes reversed.)

In Fig. 7-8(b) is shown a two-input transistor *or* circuit, though, as

194 / Integrated Circuits (ICs) and Modules

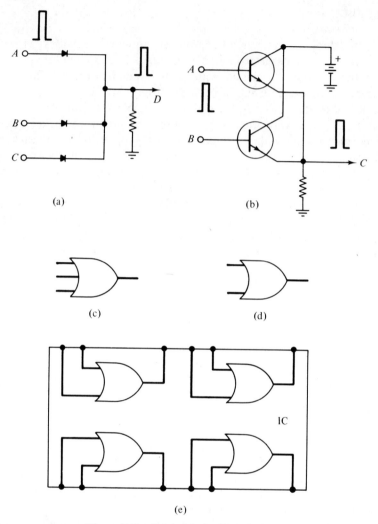

Figure 7-8 *Or* circuits and symbols

with the type shown in (a), additional transistors could be employed for multiple inputs. The transistor *or* circuit has the advantage of providing signal gain. As shown, an emitter-follower system is employed; hence, the output signal has the same polarity as the input signal.

For the circuit in (b) there is no forward bias applied, though the reverse bias is present at the collector system. The collectors are common for both transistors, and the power system would usually be bypassed to prevent any signal variations across it. The emitter terminals are also common, as shown. In the absence of an input signal, the lack of forward bias would inhibit conduction across the output resistor. Upon the application of a positive pulse signal at the *A* input, however, this transistor would have the necessary forward bias applied for the *npn* transistor used. Hence, this transistor conducts, and the current flow through the output resistor develops a voltage drop corresponding to the output signal. Similarly, a signal applied at *B* produces an output.

The symbols for the *or* circuit are shown in Fig. 7-8(c) and (d). The number of input lines present is indicated at the left of a symbol and the output line at the right. When one or more such *or* circuits are placed in an IC the representation is as shown in Fig. 7-8(e). Here, four independent *or* gates are illustrated.

Boolean algebra is the math that applies to the logic of switching and also illustrates the relationships between switching systems. In this algebra, the plus sign is used as the logical connective to denote the *or* function. Thus, $A + B$ actually denotes *A or B* and does not indicate arithmetical addition.

Truth tables are utilized to show the relationships for various combinations of inputs to switching circuits. Several symbols can be used to illustrate the logic functions. For instance, as shown in Table 7-1, the letters can be shown for switching inputs, and the zero for lack of input. Thus, Table 7-1 indicates that an output will be obtained for any combination of inputs. The true and false (T and F) can be used as shown in Table 7-2 with "true" representing an input pulse and "false" representing the absence of a pulse. Again, a true output T is obtained for every input combination except $F + F$.

TABLE 7-1	TABLE 7-2
$0 + 0 = 0$	$F + F = F$
$A + 0 = C$	$T + F = T$
$0 + B = C$	$F + T = T$
$A + B = C$	$T + T = T$

A more representative method is to use zero and one to illustrate the true and false conditions. Thus, a one would represent an input pulse and zero the absence of a pulse. For a two-input switching system for the *or* logic, Table 7-3 shows the logic. Table 7-4 shows the logic that applies to a three-input *or* circuit. These tables apply to either the diode- or transistor-type gates; they simply show the logic function and are not related to the actual circuit configuration.

196 / Integrated Circuits (ICs) and Modules

When the *or* circuit shown in Fig. 7-8(b) is altered so that the output is obtained from the collectors, the circuit appears as shown in Fig. 7-9(a). Now, the signal undergoes a 180-degree phase inversion and when an input pulse has a positive polarity, the output is negative, as shown. This inversion circuit is called a *nor*, which term is derived from the combination of *not-or* to indicate that the phase is not the same but that it is still an *or* circuit. The symbol for that *nor* circuit is shown in (c).

TABLE 7-3		TABLE 7-4		
A	B	A	B	C
0 + 0 = 0		0 + 0 + 0 = 0		
0 + 1 = 0		1 + 0 + 0 = 1		
1 + 0 = 0		0 + 1 + 0 = 1		
1 + 1 = 1		0 + 0 + 1 = 1		
		1 + 1 + 0 = 1		
		0 + 1 + 1 = 1		
		1 + 0 + 1 = 1		
		1 + 1 + 1 = 1		

7-7 AND GATE CIRCUITRY

In all logic-gate circuits, design factors are related to the nature of the signals utilized. In some digital logic applications a negative-polarity signal is considered logic one, and the absence of a signal is considered logic zero. On the other hand, some engineers may recognize a positive signal as logic one and the absence of a signal as zero for their particular design and systems.

Another type of logic-gate circuit is shown in Fig. 7-9(b). Here, positive pulses are again applied and *npn* transistors used for simplicity in integrated circuit design. For this circuit, note that the emitter of the upper transistor is in series with the collector of the lower. Thus, no current flows through the output resistor unless *both* Q_1 and Q_2 transistors conduct simultaneously. Thus, if a pulse is applied at one of the inputs, the positive polarity would supply the necessary forward bias to that transistor, but the absence of a pulse on the other transistor would prohibit conduction, since an open series circuit persists. If, however, a pulse is applied to each input simultaneously, conduction will occur, because both transistors have the necessary forward bias applied. Again, an emitter-follower circuit is used, and consequently the output pulse has the same polarity as the input pulse.

The term *gate* is applicable to this circuit, because the input for either one transistor or the other can be considered a gating signal. Thus, if a signal is applied to the upper transistor, no output will result. With the signal applied to one transistor, however, a coinciding signal at the other input opens the circuit gate, and an output signal is obtained. The term *switch* also applies, since one input can be considered as a switch function for the other to close (theoretically) the circuit and produce an output signal. Such a circuit is an *and* type and is sometimes called a *coincidence gate*.

The *and* function in logic expressions is indicated by the multiplication sign to represent the logical connective. Thus $A \cdot B$ indicates A and B, but does not mean the arithmetical multiplication of one by another. The logical connective dot could be omitted and the common indication for algebraic multiplication used by placing the letters close together, as AB. Hence, for a two-input *and* gate the logic expressions for every possible input combination, in truth-table form, are

TABLE 7-5

$0 \cdot 0 = 0$	$0 \cdot B = 0$
$A \cdot 0 = 0$	$A \cdot B = C$

Thus, an *and* circuit produces an output only if both inputs are present, that is, A and B.

The symbol for the *and* gate is shown in Fig. 7-9(d). If the output for the circuit in (b) is obtained from the collector, the consequent phase inversion of the signal produces what is known as a *nand* circuit (*not-and*). The symbol would then be as shown in Fig. 7-9(e).

Additional inputs could, of course, also be included for the circuit in (b) by adding additional transistors as well as the accompanying input resistors. For a three-input *and* circuit, if logic 1 and 0 symbols are used, the truth table is

TABLE 7-6

$A \cdot B \cdot C$			Output	$A \cdot B \cdot C$			Output
0	0	0	0	1	1	0	0
1	0	0	0	1	0	1	0
0	1	0	0	0	1	1	0
0	0	1	0	1	1	1	1

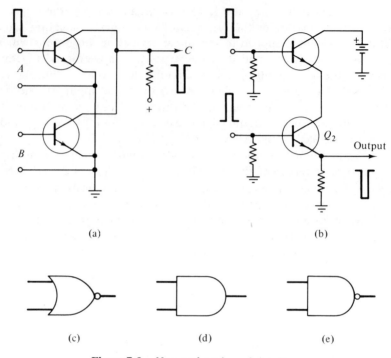

Figure 7-9 *Nor, and,* and *nand* circuits

Thus, such an *and* gate has an output only when there is an input applied simultaneously to all three inputs. For any two inputs, no output is obtained.

7-8 LOGIC CIRCUIT COMBINATIONS

Logic circuits can be combined in a number of ways to perform specific tasks. Once the design layout has been completed, it becomes part of the integrated circuit, as shown in Fig. 7-10(a). Here, the two *and* circuits feed an *or* circuit. Thus, the IC has four inputs and one output, with each *and* gate requiring simultaneous signal inputs to produce an output.

Interconnected *or-and* circuits are shown in (b). Here, two transistors are operated in series; these are in parallel with a single transistor, as shown. Thus, the *or* and *and* functions are combined, and because the output is obtained from the collector side, the output signal is inverted. Thus, the *or* and *and* functions become *nor* and *nand*. To show such negated functions, an overbar is used to express the letter symbols as having been inverted. As shown in (b), the output then consists of an

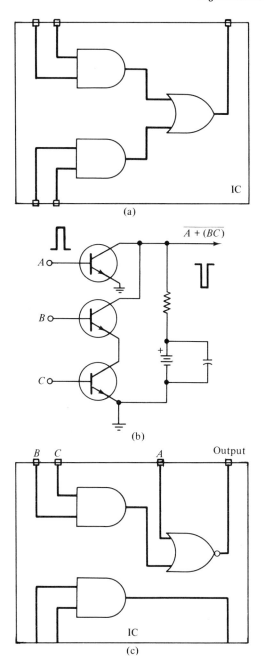

Figure 7-10 Logic circuit combinations

inverted A, which is the *or* function, plus the inverted B and C outputs, which form the *and* functions. Thus, the output expression is

$$\overline{A} + \overline{(B \cdot C)}$$

The symbol representation for the circuit in Fig. 7-10(b) is shown within the integrated circuit in (c). For this IC there is also a single *and* circuit provided, as shown. As noted in (c), the *or* circuit has an A input, while the *and* circuit has the B and C inputs. Both the B and C inputs must be present to appear at the output, though A need not be present for an output from the *or* circuit.

The circuit shown in Fig. 7-10(b) is sometimes called a *transistor-transistor logic* circuit; the symbol for this is TTL. The more general TTL is covered in Sec. 7-11.

Single-transistor logic gates can also be formed by using input resistors in a voltage-divider arrangement, as shown in Fig. 7-11(a). A

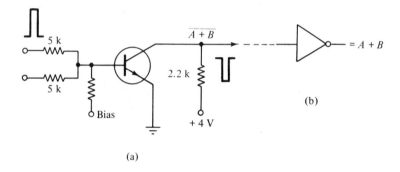

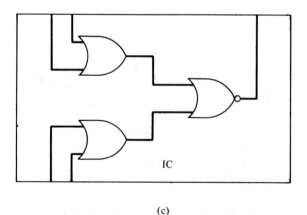

Figure 7-11 Resistor-transistor logic (RTL)

multiple-input circuit of this type can also be designed. For the circuit shown, an inverted gate function (*nor*) is obtained, though *and* gates with multiple-resistor input lines could also be assembled. For the circuit shown in (a), reverse bias instead of forward bias is applied to the bias resistor feeding the base of the transistor. Such reverse bias, when present at both the base-emitter circuit and the collector-emitter, prevents the transistor from conducting. If a positive-pulse voltage of sufficient amplitude is applied to either or both the *A* and *B* inputs, the reverse bias between base and emitter is overcome and the transistor conducts, producing the inverted output in typical *nor* function. If this circuit is followed by a standard amplifier that provides a phase reversal, the *or* function is regained at the output of the amplifier. Such a phase-inverting amplifier is often termed a *not* circuit, because the output phase is not the same as the input signal. The *not* circuit has for its logical symbol the standard triangle indicating the amplifier, followed by a small circle to indicate the inversion process, as shown in Fig. 7-11(b).

Figure 7-11(c) shows another IC system, wherein two *or* circuits have an input supplied to a *nor* circuit as shown.

For the circuit shown in Fig. 7-11 the resistor-transistor logic is designated by the symbol RTL.

7-9 DIODE-TRANSISTOR (DTL)

Diodes are also used for the inputs to logic circuits, as shown in Fig. 7-12. Here, an emitter-follower transistor is again used, though an amplifier of the ground-emitter type could also be used if the signal voltage gain is needed and the *not* function is acceptable. For the circuit shown, a reverse bias potential is applied to the bias resistor feeding the base, and hence operation is beyond the transistor cutoff point. For any input signal having a positive polarity sufficiently high to over-

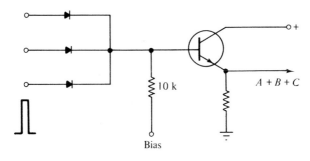

Figure 7-12 Diode-transistor logic (DTL)

come the reverse bias, conduction will occur, and an output signal will be produced having the same phase as the input signal. Thus, the $A + B + C$ logic is obtained with this circuit. In similar fashion, *and*-gate circuits using diode-transistor combinations can be formed.

7-10 LOGIC POLARITY FACTORS

Binary (base 2) notation is utilized in switching systems, as previously mentioned. Thus, only the zero and one numerals are utilized. Computers use the binary equivalents of the base ten numbers, as shown in the short listing given in Table 7-7.

TABLE 7-7

Base 10	Binary	Base 10	Binary
0	0000	6	0110
1	0001	7	0111
2	0010	8	1000
3	0011	9	1001
4	0100	10	1010
5	0101	11	1011
	etc.		

Note that in the first-place column for the binary numbers, 1 alternates with 0 for the entire vertical column at the right. In the second-place column, two 0's alternate with two 1's in progression down this column. In the third-place column, four 0's alternate with four 1's for the entire length. If the binary numbers up to 100 or 200 were to be included, this same radix-2 factor would prevail for all the successive columns added to the left.

The binary system is based on the powers of 2, where the first place has a power of 2^0, the second-place numeral has a power of 2^1, the third has a power of 2^3, etc. The place relationships with respect to value are as shown in Table 7-8.

TABLE 7-8

Place:	8	7	6	5	4	3	2	1
Value:	128	64	32	16	8	4	2	1
Hence:	. . . .			1	0	1	1	0 = 16 + 4 + 2 = 22

Table 7-8 is useful for ascertaining the decimal equivalent of a binary number. As shown, for the binary 10110 the base-10 equivalent is 22.

Computers utilize binary numbers to perform all the arithmetical functions that we perform with the base-10 system. From the standpoint of utilizing pulse signals to represent binary numbers, the relation of logic 1 or logic 0 to signal voltage polarity must be considered. One common usage is to assign a positive voltage to represent logic 1 and a negative voltage to represent logic 0. In this case, the positive and negative states are *relative to each other* and are not necessarily positive or negative with respect to circuit ground. Thus, a positive voltage may represent 1 and a less positive voltage may indicate 0. Since the 0-voltage representation is less positive, *it is negative with respect to* the logic 1 signal in this instance.

When a positive signal represents 1, it has been called the *up-state signal* or *up-level signal,* and the negative signal for 0 has been termed the *down-state* or *down-level signal.* When a negative signal is used to represent logic 1, a less negative signal (or a 0 signal) indicates logic 0. Again, a less negative signal for 0 is an equivalent positive signal with respect to that for logic 1.

The *not* function illustrated earlier in Fig. 7-11(b) must also be understood, because it has no equivalent in ordinary algebra. As mentioned earlier, an overbar over an A, for instance, means *not A*, and similarly if an overbar were placed over 1, it would represent *not* 1. Because an overbar over 1 indicates a negated 1, it must have a value of 0, since there are only two variables in the binary notation. Similarly, $\overline{0}$ has a value of 1, and if A is 1, then $\overline{A}$ (*not A*) must represent 0. If, however, A is assigned a value of 0, then $\overline{A}$ is 1. Such letter symbols assigned to the input lines of logic gates and switches are not confined to A and B but depend on the logic involved in the particular design representation to be indicated.

A double negation converts an expression back to its original form: $\overline{\overline{A}} = A$. A statement such as $\overline{A} + A$ is a true statement, since it indicates that either one or the other is to be considered. The statement $\overline{A} \cdot A$, however, is a false statement, because the combination of the negated A and the unnegated A (a negative and positive value) cancels out.

7-11 MULTIEMITTER TTL

A transistor with more than one emitter is often included in ICs for forming logic gates. Extended emitter-base diffusion processes produce two or more emitters having a symbol representation as shown in Fig. 7-13(a). This unit is useful for multiple-input logic-gate circuitry and lends itself to integrated circuitry because of its basic simplicity and ease of formation.

Utilization of the multiemitter transistor to form a logic gate is shown in (b). Here, a three-emitter *npn* transistor is used, with emitters forming three inputs as shown. Transistor Q_2 is an inverter, thus forming a negated logic for the Q_1 and Q_2 combination. This system forms a *nand* gate, and the logic is transistor-transistor (TTL).

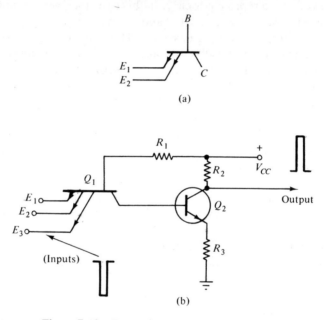

Figure 7-13 TTL using multiemitter transistors

Since both transistors are *npn* types, forward bias consists of a positive base with respect to a negative emitter. Hence, for Q_1, there is no conduction until a negative (forward) voltage is applied. Thus, negative pulses are used to trigger the system.

The collector for Q_1 is directly coupled to the base of Q_2, as shown. Any signal applied to the base of Q_2 undergoes a phase inversion when it appears at the collector output (junction of R_2 and collector). A negative pulse applied to any of the emitter inputs is ineffectual in altering the conduction of Q_2. With the application of three negative pulses, however, the in-phase potential change at the collector of Q_1 is felt at the base of Q_2, and conduction declines. Thus, the voltage drop across R_2 decreases and the collector voltage rises toward the positive supply potential, producing a positive pulse output as shown.

As with typical IC formation, a minimum of components is involved with the multiemitter transistor and the TTL circuitry. Direct coupling is utilized between the two, thus eliminating a coupling capacitor and its associated disadvantages in attenuating lower frequency signals. The switching response is very rapid, because the multiemitter transistor has negligible delaying effects on gating characteristics when Q_2 conducts. More than three multiemitter inputs can be used (the term *fan-in* is sometimes used to denote the input segments), but noise problems arise and efficiency declines materially.

7-12 MOSFET NAND

The MOSFET also lends itself to the formation of logic gates; two examples are shown in Fig. 7-14. Both of these are *nand* circuits, because there is the usual 180-degree phase inversion from input to output of the grounded-source transistor.

A four-input MOSFET circuit is shown in (a), with steering diodes forming a one-way current path. When four inputs are applied simultaneously, the drain-to-source current change produces an inverted signal as shown, forming the logic expression given at the output. Such a circuit, using diodes and transistors, is sometimes referred to as a DTL (or TDL) gate.

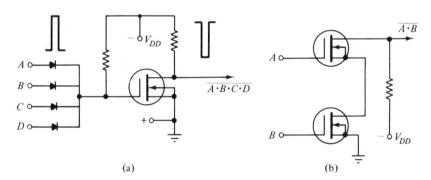

Figure 7-14 MOSFET *nand* circuits

In Fig. 7-14(b) an alternate arrangement is shown, using two MOSFETs. Here, circuit bias is applied to inhibit conduction for either transistor. When a pulse is applied at A the associated transistor can conduct, but since the current path is in series with the other transistor, no conduction occurs unless inputs are applied to A and B simultaneously. Again, a negated output occurs as shown.

7-13 DCTL CIRCUITRY

A representative direct-coupled transistor logic (DCTL) circuit is shown in Fig. 7-15(a). Note the simplicity of these multigate transistor circuits, where only three transistors are required to perform the logic functions. Thus, such a circuit again lends itself readily to IC processing.

The first transistors, in parallel, perform an *or* circuit, as shown symbolically in (b). Thus, an input at either one or more of the base elements for the initial group of three will cause conduction, because a positive pulse simulates the required forward bias. When any one or more of these transistors conduct, they form a virtual short circuit across the output, since impedance is extremely low during conduction. Also, the current flow through R_1 causes a large voltage drop to occur, placing the junction designated as (x) at virtual ground, or negative polarity. When this occurs, the base inputs of the succeeding *or* circuits have reduced or zero bias applied, and conduction ceases. Now, the points marked (y) and (z) have a polarity and potential equal to that of the source, since very little voltage drops across resistors R_2 and R_3. Thus, a positive input pulse at the first *or* circuit produces a positive voltage change at outputs marked O_1 and O_2; hence the output phase is the same as the input. This comes about because of the phase inversion for the *or* circuit between base and collector. Thus, as shown in (b), the output from the first *or* circuit is negated, and when this is applied to the inputs of the succeeding *or* circuits, the signal is negated again and is brought back to its original polarity.

For the additional inputs at the second and third *or* circuits, an input would be inverted; the inversion produces a negated output.

7-14 COMMERCIAL IC PACKAGES

In addition to the common and general circuits, such as amplifiers, oscillators, switches, and gates, a number of special IC packages have been designed and produced by electronic manufacturers. Such packages contain specific features and advantages that make them attractive for inclusion in planned (or existing) electronic systems and thus expedite design. Often such special IC units contain only a few transistors in a certain circuit arrangement that makes for low cost but high efficiency and performance. Other ICs may combine such units as Schottky diodes (see Sec. 1-6) with TTL circuits for producing a special switching system having superior gating characteristics in terms of speed.

Virtually all the basic circuits discussed in previous chapters can be found in integrated circuit networks, and individual units of special

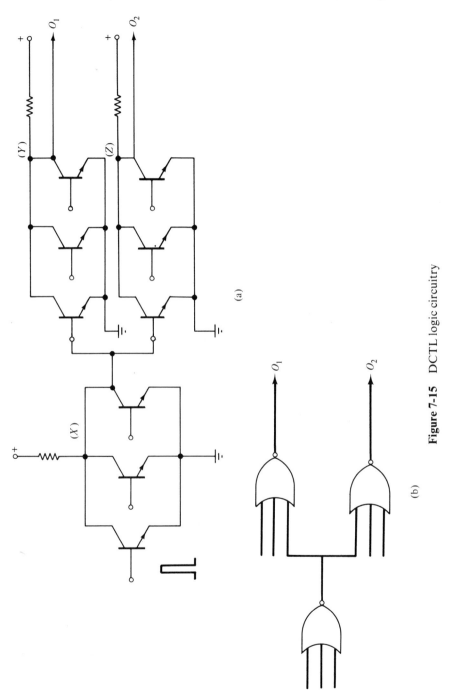

Figure 7-15 DCTL logic circuitry

merit (such as the Schottky diode) are included in specific commercial packaging as required. On occasion only special-type FET units may be packaged, or combined with differential or operational amplifiers to perform some specific task. In succeeding sections of this chapter the characteristics and qualities contained in some of the special package units are discussed. Thus, their particular features can be understood when such may be needed in the assembly and design of some particular electronic system.

7-15 C-MOS UNITS

Figure 7-16(a) shows the basic makeup of the C-MOS device originally pioneered by RCA. The term stems from the fact that it contains a *complementary* circuit combining two different-channel MOSFETs as shown. The C-MOS has a number of advantages in IC technology, and can be used for both digital and linear systems.

The C-MOS consumes virtually no power in the absence of applied signals, and a logic array made up of such units with as many as 100 gate circuits may consume less than 0.1 mW of power. The low-power operation is possible because of the *p*-channel and *n*-channel MOSFETs connected in parallel as shown. Thus, because of opposite-polarity characteristics, if the *p*-channel device is gated into conduction, the *n*-channel unit will be in a nonconducting state. Consequently, the

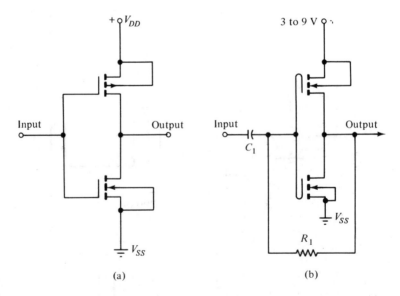

Figure 7-16 C-MOS circuitry

operational currents are extremely low. During operation in digital circuitry, where the signals consist of short-duration pulses, each transistor operates intermittently, thus again keeping current consumption at extremely low levels. As the switching speed is increased, the average power also increases, though if gating speeds up to 10 kHz are used, the unit dissipates less than a microwatt for any single gate. As frequencies are increased, the repetition rate means that the current pulses occur at a shorter interval of time, and consequently power consumption increases to some degree.

The logic-transfer characteristics are exceptionally excellent, so the C-MOS has a high noise immunity. A number of manufacturers are now producing digital C-MOS units capable of operation in the megahertz region. Hence, in addition to the usage of C-MOS in computating and calculating systems, they are also of value to the designers of communication and instrumentation equipment.

For linear-circuit operation a coupling resistor (R_1) may be added between the input and output to stabilize the circuit bias, as shown in Fig. 7-16(b). A feedback path is formed by R_1 that corrects for any drift sensed at the output terminal. The feedback resistor stabilizes the bias with virtually no voltage drop across R_1 (because of negligible gate current flow during normal operation). Hence, R_1 can have an ohmic value in excess of 20 MΩ for effective signal (ac) isolation.

The circuit in (b) can be a foundation unit for amplifiers or oscillators. For crystal oscillators, the crystal is put in shunt with R_1 and forms an equivalent resonant-circuit feedback loop for the signals to sustain oscillations and maintain close frequency stabilization.

Certain terms and symbols have been accepted as standard among C-MOS manufacturers. The more general ones are listed in Table 7-9.

TABLE 7-9

Symbol	Representation	Explanation
V_{DD}	Drain voltage (high)	Most positive voltage applied to unit, such as + 9 V
V_{SS}	Source voltage	Most negative supply potential applied and used as reference for other voltages (generally indicative of ground reference)
V_{IH}	Input voltage (high)	High–logic level range of input V
V_{IL}	Input voltage (low)	Low–logic level range of input V

(Continued)

TABLE 7-9 *(Continued)*

Symbol	Representation	Explanation
$V_{IH(min)}$	Minimum input voltage (high)	Minimum permitted input high-logic level
$V_{IL(max)}$	Maximum input voltage (low)	Maximum permitted input low-logic level
V_{OH}	Output voltage (high)	Range of voltages at output terminal with specified output load and given supply potential. (Inputs conditioned for high–logic level at output)
V_{OL}	Output voltage (low)	Range of voltages at output terminal with specified output load and a given value of supply potential. (Unit inputs are set for low-logic level at output)
I_{IN}	Input current	Current flow into unit at a given input voltage and V_{DD}
I_{OH}	Output current (high)	Drive flow out of unit at given output voltage (logic high) and V_{DD}
I_{OL}	Output current (low)	Drive current flow into unit at given output E (logic low) and V_{DD}
I_{DD}	Quiescent value of supply source I	Current flow into drain terminal at given input signal and established V_{DD} values
t_{TLH}	Transition time, low to high	Time between two specified reference points (usually given between 10 and 90 percent) on a waveform change from low to high
t_{THL}	Transition time, high to low	Time between two specified reference points (usually 90 and 10 percent) on a waveform changing from high to low

7-16 INTEGRATED INJECTION LOGIC

One of the most important and significant electronic innovations for ICs in recent years has been the design and production of an integrated-circuit chip known as *integrated injection logic*. The symbol for this type of circuit is the unusual one designated as I^2L.

The I^2L system is characterized by high performance of bipolar LSI circuits, very low power consumption, a highly reduced-size chip compared to other logic bipolar chip design in use before its invention, and the simplicity of design owing to the minimum amount of associated components. Consequently, as many as 3000 gates or switches are present in a single chip, or for binary-signal storage, 10,000 bits of memory are easily possible.

Despite the number of gates that may be incorporated in a single chip, the power dissipation is less than that of several hundred gate units used with older chip processing. Bipolar logic is simplified with the I^2L design, since it provides for low cost and utilizes conventional processes (bipolar) in microminiaturization.

The versatility of the I^2L chip extends into many areas of electronics. One application is the production of low-cost chips for electronic watch circuitry, where only microwatts of energy are dissipated while direct high-current drive signals are still provided for actuating the numeric displays of light-emitting devices. In addition, the I^2L chips are useful in digital voltmeter circuitry, digital tuners, frequency division of signals generated by basic oscillators such as those used in electronic organs, as well as in linear circuitry for radio and television. In addition, the I^2L chips are useful in logic arrays, read-only memories, and the circuitry for the production of the logic for calculators, including shift registers and signal conversion.

Basically, the I^2L circuit forms a gate by utilizing a pair of transistors having complementary characteristics. An *npn* transistor with multiple collectors is utilized as an inverter, and a *pnp* transistor behaves as a load as well as a current source. (No resistors are utilized for either the load or source functions.)

The basic unit is illustrated in Fig. 7-17(a). Here a constant-current source (see Sec. 6-15) is used with a multicollector *npn* transistor as shown. Using a *pnp* transistor as the constant current source, we obtain the final form of the I^2L unit, as shown in (b). Now the first transistor (*pnp*) serves as a direct-coupled gate injector to the base of the second (*npn*) transistor as well as an inverter. The second transistor is the load and eliminates the conventional resistor load. Compare this multicollector configuration with the multiemitter TTL circuit shown in Fig. 7-13.

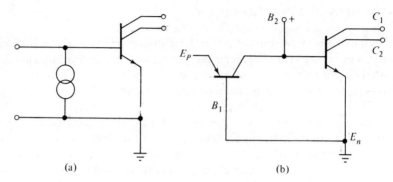

Figure 7-17 I^2L circuitry

In the formation of the I^2L chip both the circuit elements utilize the space of a single IC transistor; thus there is no need for device isolation, as there is with other chips. It is for this reason that the fabrication of the I^2L chip presents no more problems than that of a single planer transistor.

Basically, the fabrication of the I^2L consists of utilizing an n-type substrate for the base of the device. The substrate also serves as a common ground plane for interconnection of the grounded-emitter gates, as shown in Fig. 7-18. This process eliminates the need for ground interconnections and adds to the simplicity of the design. An n-type epitaxial layer is grown over the substrate, and this layer serves as a grounded-emitter section of the npn switch as well as the grounded-base section of the pnp injector transistor. Two diffusions are utilized, the first forming the p-base of the npn, as well as the p-collector of the injector transistor. The second diffusion completes the I^2L gate by furnishing a multiple-collector series of regions of the npn.

The regions marked $n+$ are those where heavy n doping is used for a high-concentration layer as opposed to regions that are designated as

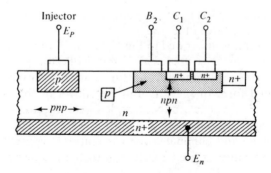

Figure 7-18 I^2L structure

n, or lightly doped $n-$. [Such markings are also used for p doping ($p+$ and $p-$).]

Etching of metal deposits is then performed to provide the necessary interconnections between the various transistor elements. In essence, the I^2L multicollector sections parallel in design the TTL's multiemitters, as was shown in Fig. 7-13. Similarly, the I^2L and TTL occupy (insofar as the bases are concerned) similar areas. Also, the I^2L emitter and the TTL collector are situated at the lowest-end regions of the IC. Basically, the transistor logic in I^2L is obtained by regulating the inverse beta, while in TTL the forward beta is under control.

The I^2L performance under various types of signals is particularly stable, and there is no increase in the power utilized for an increase in the frequency of the signals handled. Also, the circuit is particularly immune to switching noise and transients.

Injection logic can be compared to the direct-coupled transistor logic (DCTL), as shown in Fig. 7-15. For the I^2L, however, the different bases and emitters are combined into a single multicollector transistor by the diffusion process. Virtually any standard bipolar process utilized in IC fabrication can be employed for the formation of the circuit elements of the I^2L system.

7-17 I^2L NOR CIRCUIT

A two-input *nor* circuit with an additional *or* output is shown in Fig. 7-19. Here, three I^2L units are used, and the constant current sources are formed by the injector transistors, which are a part of the I^2L system.

Note that the collectors of each input I^2L unit are cross coupled, so that the I^2L unit with input A has the upper collector connected to the lower collector of the I^2L unit with the B input, etc. For the A input, the output is fed to an additional I^2L device for phase inversion purposes. When a signal is injected into the A input, it undergoes a phase reversal at the output of the unit; when this is applied to the second I^2L unit, it reinverts the signal, bringing it back to the original logic expression. For the B input, however, a phase inversion occurs across the transistor, and hence the output is an inverted form.

If an input is applied at A for the upper network, it will be applied to the base of the invertor I^2L unit and hence will appear at the output as an A output. The signal from the first I^2L unit, however, is also applied to the lower collector output, and, since it is not reinverted, it appears at the output as a negated A.

Similarly, for an input pulse at B, the output from the lower collector would be an inverted B, as shown. The output from the other

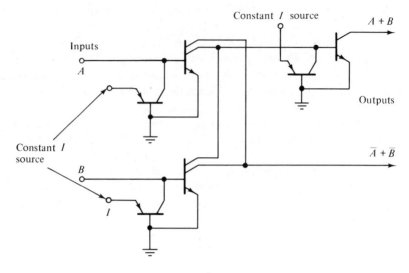

Figure 7-19 I^2L *nor* circuit

collector of the B input is applied to the upper circuitry, and to the output invertor as shown. Consequently, the B appears at the output line as a noninverted logic, as shown. Thus, an output pulse is obtained for either an A input *or* a B input *or* both, and an output is obtained in a noninverted form as well as in a negated form as shown. By adding other circuits similar to the ones shown in Fig. 7-19, additional inputs or outputs can be obtained as required.

7-18 SCHOTTKY CLAMP FOR I^2L

The Schottky diodes discussed earlier in Sec. 1-6 are useful for aiding the performance of other IC circuitry. One application is shown in Fig. 7-20(a), where such special diodes are used to clamp the output signals of an I^2L gate circuit. The Schottky diodes limit the swing of the logic signals, and because such signal amplitude limitations are imposed, there is a reduction in switching delays, which would otherwise be encountered for excessive excursions of the signal swing. The utilizations of the Schottky diodes in this manner can increase the gating and switching speeds of the I^2L circuits up to five or six times that of the I^2L gates not using the diode clamping processes.

The circuit shown in Fig. 7-20(a) is that used by International Business Machines Corp. for their I^2L integrated circuitry to reduce the amplitude swing of logic signals from over 500 mV to a much lower

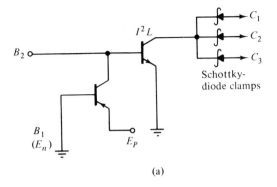

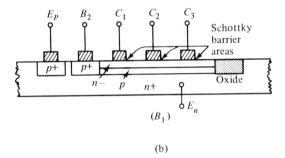

Figure 7-20 Schottky-clamped output for I^2L

range lying between 150 to approximately 300 mV. Consequently, switching speeds are increased to the degree mentioned earlier.

The basic structure for the I^2L, which also incorporates the Schottky diodes in fabrication, is shown in (b). Thus, the diode clamps are built into the IC, and usually the process employs only an additional masking procedure. The identification symbols shown in (b) can be compared to those given in (a) to show the relative disposition of the I^2L and Schottky diodes in the IC structure.

7-19 PACKAGED ICs

In addition to the commercially available packaged ICs containing one or more units such as C-MOS, I^2L, Darlingtons, and others, a variety of other devices are included in integrated circuitry, many of them containing conventional amplifiers, demodulators, and signal processing circuits. Often the C-MOS or I^2L devices may be combined with junction transistors to form a hybrid IC for performing some logic function or processing a signal in a linear amplifier. Thus, the design

216 / Integrated Circuits (ICs) and Modules

engineer must become acquainted with the units generally available for combination in a complete system under design.

Often an IC package will be designated as containing several flip-flop circuits, and the circuitry will not be shown in the manufacturer's literature, but only the logical sequences utilized. Also, circuits involving mathematical functions or gating and switching combinations for control may also be shown only in a logical sequence or logic symbol form without indicating particular circuitry. Typical is the half-adder shown in Fig. 7-21. Basically the half-adder consists of an *or* circuit, an *and* circuit, and another *and* having a negated input as shown (the latter forming what is known as an inhibiting system). These three logic symbols may be shown in the IC identification illustration as in (a). Often the single symbol shown in (b) is used to indicate the same circuits without the carry line and termed an *exclusive-or* circuit. If the circuit shown in (b) is followed by an inverter [symbol as at (c)], it becomes an exclusive *nor* circuit, and the symbol would then be as shown in (d).

For the half-adder shown in (a), serial-train full adders can be assembled that use two of the half-adder circuits. The half-adder is also useful for other switching purposes and for code conversions.

For the inhibiting *and* shown at the output in (a), a steady-state voltage may be applied to the internal circuitry so that the inhibiting input is normally in the *on* mode, thus permitting the entry and output

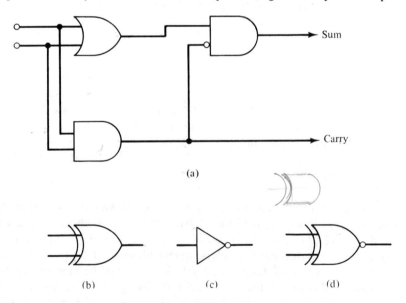

Figure 7-21 Exclusive-*or* (half-adder)

of any number obtained from the upper *or* circuit. When, however, two pulses are applied to the *or* circuit inputs, they also appear at the lower *and* circuit. Hence, an output is obtained from the *and* circuit that appears at the inhibiting input, stops conduction of this circuit, and prevents the entry of signals from the *or* circuit. The logic, therefore, is that when a binary one appears at both *or* circuits, a zero output appears at the sum output, but a one appears at the carry output. For individual pulses applied to the *or* circuit (not simultaneously), no inhibiting pulse appears, and, therefore, the output is obtained at the sum output. This follows the binary-addition principle where one plus one equals binary 10 (base-ten 2). Thus, a 1 + 1 input would automatically cancel any output at the sum output, but would provide the carry pulse for the second-place one in the binary 10.

From the foregoing description of the logic involved, Table 7-10 can be formed to give a graphic illustration of the input versus output modes.

TABLE 7-10

A B	Output	A B	Output
0 + 0 =	0	0 + 1 =	1
1 + 0 =	1	1 + 1 =	0

An IC package can also be formed of half-adders to convert the Gray code to pure binary, as shown in Fig. 7-22. The Gray code has been widely used in computer and control systems, because operational errors are reduced. This comes about because only one digit changes at a time for the Gray code as the numerical value increases. This is not the case with pure binary, as shown by Table 7-11.

TABLE 7-11

Base Ten	Binary	Gray	Base Ten	Binary	Gray
0	0000	0000	7	0111	0100
1	0001	0001	8	1000	1100
2	0010	0011	9	1001	1101
3	0011	0010	10	1010	1111
4	0100	0110	11	1011	1110
5	0101	0111	12	1100	1010
6	0110	0101			

218 / Integrated Circuits (ICs) and Modules

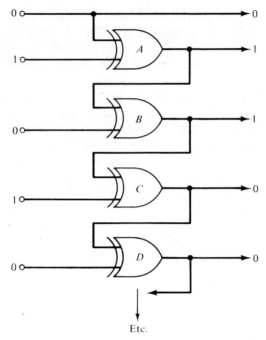

Figure 7-22 Gray code to binary

The Gray code is also called the *cyclic code,* or *minimum-error code.* Another term for this code is *reflected binary.*

For the string of *exclusive-or* circuits shown in Fig. 7-22, as many are used as necessary for the length of the numbers to be handled. As an illustration of its function, assume that the Gray code number 1010 is entered at the left as shown. Remember that the half-adder performs the addition of 1 + 1 *without carry.* The function is as follows: The 0 entered at the upper input of the *A or* gate also appears at the output, since it is a continuous unbroken circuit as shown. For the 1 entered at the lower input of *A*, a 1 will be produced as shown at the output. This is also recirculated and applied to the input for the *B or* circuit. Since this *or* circuit has no input for the lower gate, a 1 appears at the output. This is recirculated into the upper gate input for *C*, but since the second gate for this *or* circuit also has an input, no output is produced, as shown. Similarly, for the *or* circuit at *D*, no input is present at either gate; hence no output is obtained. Thus, the Gray code number 1010 has been converted to the binary code number 1100, as may be seen by reference to these numbers in Table 7-11.

In computers, complement numbers of the binary representations are often used, such as 0101 as the complement of 1010, or 1010 as the complement of 0101, etc. The complement consists of replacing 1's with 0's, and 0's with 1's.

A series of *and* circuits and *or* circuits can be combined to provide for selection of either the pure binary number or its complement. Typical of an IC package is the representation shown in Fig. 7-23. The steady-state values of flip-flop circuits (see Fig. 6-1 and related discussions) are used to apply a potential to the *and* gates from both the 0 and 1 output lines as shown. Hence, if a flip-flop stage is in its zero state, the voltage from the 0 lines coincides in polarity with that of the complement trigger pulse, and the *and* circuit thus produces an output pulse representing logic *one* (the complement of zero). If the flip-flop circuit is in the *one* state (an interchange of the 1's and 0's shown) no polarity coincidence occurs for the complement pulse, and a zero output is produced (the complement of one).

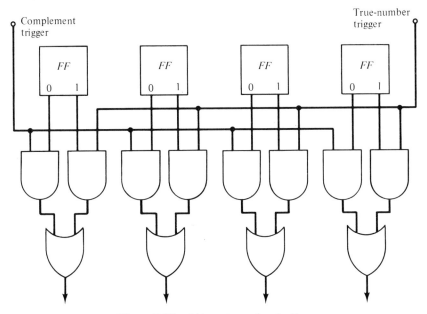

Figure 7-23 Alternate mode selection

Similarly, if a flip-flop is in its *zero* state, the voltage from the right FF line is opposite in polarity to the true-number trigger pulse, and no coincidence occurs; hence a *zero* output is obtained upon the application of such a trigger pulse to the true-number line input. If the flip-flop is in the *one* state, the right FF line has a polarity coinciding with that of the true-number trigger pulse. When the latter is applied, the *and* circuit passes the pulse and produces a *one* output from the lower terminals of the *or* circuits.

7-20 SWITCHING DESIGN PREREQUISITES

As mentioned in Sec. 7-19, there are many ICs available to help the designer to assemble specific systems as needed. If devices are to be formulated for new IC manufacture, the conventional circuitry discussed earlier using junction transistors or FETs must be used as the basic unit, unless the design engineer has innovative units original with him that are to be converted to IC. Many commercial logic systems are highly complex, and signal switching, routing, and rerouting become the prime objectives. In this area the design engineer is concerned with only a relatively few different types of circuits, but is faced with the problem of interconnection, selection, and sequencing.

The logic switching linkage of specific gates entails design procedures that are highly specialized and that require a thorough familiarization and study of switching design techniques. Of prime interest is the *simplification* of the switching sequences that are initially formulated. As an example, in Fig. 7-24 a six-input system is shown in (a) with an output of $(A + B)(A + C)(A + D)$. This can be reduced to the simpler expression $A + (BCD)$ and thus can eliminate two of

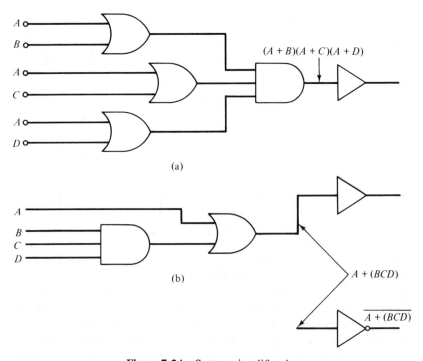

Figure 7-24 System simplification

the *or* circuits. The same logic function is still performed after circuit simplification, as shown in (b).

The example shown in Fig. 7-24 is much more simple than many encountered in practice. While the circuit in (a) with the additional switches and gates will function properly, design should be such that a minimum number of circuits are employed to save costs and fabrication time, and to produce systems that consume less power. Thus, the designer of switching circuitry networks should be thoroughly familiar with Boolean algebra, and must also have a knowledge of sequential circuit synthesis, as well as experience in the utilization of logic maps.

In many instances much loss of time and extended design practices can be avoided by using computers specifically programmed to aid in design practices for particular systems. Computer-aided design expedites much of the routine and time-consuming trial-and-error techniques sometimes necessary to obtain optimum results. The computer cuts through the number of workable designs that exist for a particular system, at the same time suggesting solutions that normally would have taken too much time to arrive at by ordinary means. Thus, "satisfactory" compromises in switching design are often avoided in favor of the superior end result. Here again, however, the design engineer must become acquainted with the capabilities of the computer, its programming language, operation of display screens, and *light-pen* manipulation (where a hand-held, pointed, electronically activated "pen" is used to alter displayed information on a viewing screen that displays computer output), as well as the operational and end-result limits that can be achieved by a particular computer.

CHAPTER 8

Power Supplies

8-1 HALF-WAVE SUPPLY

Power supplies for radios, hi-fi systems, and other electronic gear utilize various methods for rectification of the ac power mains. The most simple system is the half-wave power supply illustrated in Fig. 8-1. Here a transformer is used to step up or step down the ac line voltage to that required for rectification. In some applications the ac line voltage may be rectified directly without an intervening transformer if the potential available is that required.

Diode D_1 is in series with the transformer secondary, and when a positive alternation of the ac sine wave occurs across the transformer secondary, electron flow is along the bottom line, up through resistor R_2, through resistor R_1, and through rectifier D_1 to the top of the transformer secondary. For negative alternations, however, diode D_1 does not conduct, because the voltage appearing across the secondary would be a form of reverse bias.

Filter capacitors C_1 and C_2 have a low reactance for the ripple frequency produced during the rectification, and thus tend to smooth the ripple components to produce a dc output with a minimum amount of pulsations. The filter capacitors must have a working-voltage rating sufficiently high to withstand the peaks of the voltages developed from the rectifier. Resistor R_1 is a voltage-dropping resistor whose value is selected to produce the proper output for a given load. Resistor R_2, sometimes referred to as a *bleeder* resistor, tends to stabilize the voltage output, because it represents a steady, though small, load on the power

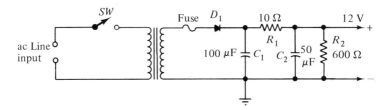

Figure 8-1 Half-wave power supply

supply regardless of the resistance variations in the actual load. (An audio amplifier, for instance, would present a varying load on the power supply, since currents would become higher during peaks of audio signals.)

The amount of current drawn by resistor R_2 depends on the voltage output. If, for instance, the output had a value of 12 V, the current flowing through R_2 ($I = E/R$) = 20 mA, since $\frac{12}{600}$ = 0.02 A.

If the load circuit connected to the power supply draws 200 mA, the total current drain from the supply (including the 0.02 mA of R_2) would be 0.22 A. Thus, the resistor R_1 must have a wattage rating sufficiently high to withstand the heat dissipation. Since $P = I^2R$, we have $0.22^2 \times 10 = 0.484$. Thus, a half-watt resistor is indicated, though a one-watt resistor would be preferable, since it operates at a cooler temperature and there is less danger of its burning out.

Approximately 2 volts develop across resistor R_1 (0.22 × 10). Hence, for a 12-V output, the secondary of the power transformer is delivering approximately 15 volts. Since this 15 V is the rms value, the peak value would be 21 V (1.41 × 15). Thus, capacitor C_1 should have a working-voltage rating in excess of this value to minimize the danger of its shorting out.

Rectifier D_1 must be capable of withstanding twice the peak voltage. This comes about because capacitor C_1 charges to the peak value, with plus toward the diode and minus to ground. When the secondary delivers a negative alternation, the transformer winding is minus at the diode, and plus at ground in series with capacitor C_1, which is minus and plus; thus diode D_1 has twice the peak voltage across it periodically, or approximately 42 volts. Diode D_1 should be capable of withstanding at least this, and consequently a rating of 50 volts or more is preferable.

Capacitor C_2 need not have as high a working voltage rating as C_1, though often both have the same ratings. For the voltages mentioned for this rectifier, a working voltage of 35 V or more can be used for C_2, the 50-V working voltage being preferable. Cost factors, of course, limit the margin of safety that it is expedient to employ.

The series fuse should have a rating of approximately 250 milliamperes to provide for a small margin of safety, and yet protect the transformer if the capacitors short or if the load on the power supply becomes excessive.

8-2 FULL-WAVE SUPPLY

A full-wave power supply is one that utilizes dual diodes, as shown in Fig. 8-2, and thus utilizes both alternations of the ac wave by having the diodes conduct alternately. For this type of power supply a transformer having a center tap is necessary; hence, the total secondary delivers twice the voltage that would be the case for the half-wave supply. As shown, if the output voltage were 60 V, then the power supply transformer secondary would have to be capable of delivering 150 V, with the center tap as shown. This is necessary so that the full 75 volts is utilized during conduction of each diode.

When the top of the secondary transformer has a positive alternation appearing on it, the bottom of the transformer secondary has a negative potential; hence diode D_2 cannot conduct. Diode D_1, however, conducts and establishes current flow through the system. Electron flow moves from the center tap through the ground wire, up through resistor R_2, through resistor R_1 (which drops the voltage to that required at the output) and returns through diode D_1 to the positive upper terminal of the transformer. For a negative alternation, the top of the transformer is negative and the bottom is positive. Now, D_1 is unable to conduct, but D_2 has the proper potentials applied for conduction. Again, electron flow is along the ground wire and through the two resistors, returning through diode D_2 to the positive lower terminal.

Voltage ratings for the capacitors should have the same considerations as those explained for the half-wave power supply. The working voltages of these capacitors should be sufficiently high that the peak voltage does not cause arcing in the capacitors and short them out.

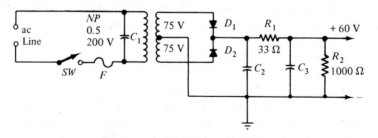

Figure 8-2 Full-wave supply

Full-Wave Supply / 225

Similarly, resistors R_1 and R_2 should have a wattage rating sufficiently high to minimize thermal damage. Again, each rectifying diode has a high voltage on it during its nonconducting state and consequently should have a sufficiently high voltage rating to minimize burnout.

For this power supply, the fuse is in the primary, and primary currents should be measured under load conditions for the power supply, so that the current rating of the fuse can be ascertained. Capacitor C_1 shunting the primary is a noise filter, and provides a low reactance for line noises. Capacitor C_1 must be of the nonpolarity (NP) type to prevent damage. (A unipolar capacitor such as the electrolytic filters cannot be used, because it is damaged by the application of a reverse voltage.) The peak line voltage (if the rms value is 120 V) is 1.41 × 120 = 169 V; hence the voltage rating of C_1 should be 200 WV or more.

The full-wave power supply is more costly than the half-wave power supply because of the necessity of a center-tap transformer and the dual diodes. The filtering efficiency, however, is superior to the half wave, and in consequence lower value filter capacitors can be used to achieve the same low-ripple output obtained for the half-wave supply with higher value capacitors. There are two reasons for this. The half-wave supply has a ripple frequency of 60 Hz, while the full-wave ripple frequency is twice that (120 Hz). A specific filter capacitor would have a reactance of only one-half the value at 120 Hz that it has at 60 Hz with consequently greater shunting of ripple components.

Another reason for the reduction of ripple amplitude is indicated in Fig. 8-3. In (a) is shown the result of half-wave rectification, where every other alternation of the ac wave is lost. Consequently, the filter capacitors charge to the peak value but must hold this charge for an interval of time before being replenished. If the current drain is sufficiently high, the discharge rate can be faster than desired, and the voltage declines appreciably before the next alternation. Thus, as at (b),

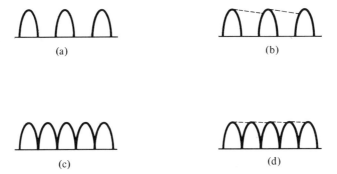

Figure 8-3 Rectification waveforms

the 60-Hz ripple is present in the dc output. For full-wave rectification, however, the missing alternations are present, as shown in (c), and every other alternation of the cycle has been inverted so that all have the same polarity. Now the valleys between peaks are much more shallow, as shown in (d), and the filter capacitors need not hold their charge for too long a period before being replenished. Hence, the ripple amplitude is much lower, and a smoother dc output is produced.

8-3 BRIDGE RECTIFIER

The bridge rectifier system provides a means for obtaining full-wave rectification without a center-tap transformer. One disadvantage with this system, however, is that four diodes are required instead of the two that were shown for the full-wave rectifier in Fig. 8-2. Another disadvantage is that the direct ground of one section of a transformer (as is the case for the half-wave and full-wave rectification) is not achieved with a bridge rectifier. Thus a greater degree of filtering may be necessary to obtain the same low value of ripple frequency as is the case of the full-wave supply of Fig. 8-2.

A typical bridge rectifier system is shown in Fig. 8-4. The same considerations apply for the filter capacitor sizes, the working voltages, and the wattage ratings of the resistors as were given for the half-wave and the full-wave rectifier systems. For the bridge system shown, two diodes conduct alternately with the other two to provide for full-wave rectification. If, for instance, a positive alternation appears across the secondary of the transformer, the top of the secondary becomes positive and the bottom negative. Consequently, diode D_3 conducts and electrons flow through it and along the ground terminal through resistor R_2 and then R_1. The return path to the top of the transformer

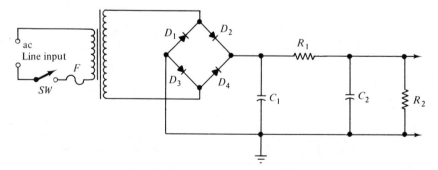

Figure 8-4 Bridge rectifier system

winding is provided by diode D_2, which, with diode D_3, finds forward-bias conditions when a positive alternation appears across the secondary.

For a negative alternation, the top of the secondary winding is negative and the bottom is positive. Now, the diode D_1 conducts, and electrons flow through it, again along the ground interconnection and through the resistors; a return path of the electrons to the positive terminal (bottom of transformer) is provided by diode D_4. Thus, full-wave rectification is provided, with each alternation of the ac signal being utilized.

8-4 VOLTAGE DOUBLING

On occasion the transformer used for stepping up voltages in power supplies can be eliminated and a higher voltage than that present in the ac line can be obtained. This is accomplished by using circuits that have the capacity for doubling or tripling existing voltages. Such rectification systems can also be utilized to obtain a higher voltage than that available from the secondary of a given transformer. The system works on the principle of charging individual capacitors to the voltage available, and then utilizing the total charge of the two capacitors in series to obtain a higher voltage.

A typical voltage-doubling system is shown in Fig. 8-5. Two diodes are used as shown, plus two associated capacitors utilized for the voltage-doubling process. For convenience in analysis, the input line terminals have been identified as T_1 and T_2.

When an input alternation of the ac cycle is such that terminal T_1 is positive and T_2 is negative, electrons flow through the T_2 line and,

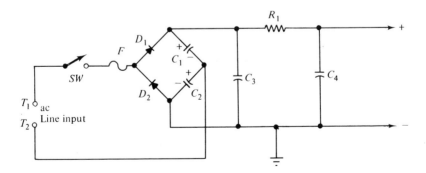

Figure 8-5 Voltage doubler

in seeking a return path to the positive terminal T_1, charge capacitor C_1. (Electrons accumulate on one side of the capacitor section and leave the other side, to flow through diode D_1 to the terminal T_1.) For the next alternation, where T_1 is negative and T_2 positive, electrons flow into the T_1 line and through diode D_2; in the return path for the electrons to terminal T_2, capacitor C_2 is charged to the peak voltage of the alternation, as is the case with capacitor C_1. Now capacitors C_1 and C_2 have been charged with a voltage polarity as shown.

As shown in Fig. 8-5, the output voltage to the filter section is obtained from the junction of D_1 and C_1 (positive output) and from the junction of D_2 and C_2 (negative output). Thus, since each capacitor is charged to the line input peak alternations, the output is the sum of the two voltages, or double the input voltage. Thus, if the line voltage is 120 V, the output voltage from the rectifier system is approximately 240 V, though under load conditions less voltage is available, the amount depending on the voltage drop across R_1 and the degree of regulation present. (Regulation is covered fully in Sec. 8-6.)

For voltage requirements above or below the ac line input, a transformer is utilized, preceding the rectifier filter section, as shown in Figs. 8-2 and 8-4.

Even though each alternation of the input ac cycle is used, the voltage-doubling capacitors receive only alternate charges, as is the case with the filter capacitors for half-wave supplies. Thus, variations in the power requirements of the load affect this circuit to a greater degree than would be the case for full-wave rectification. Capacitors C_1 and C_2 of Fig. 8-5 charge to the peak voltage of the ac line, while filter capacitors C_3 and C_4 charge to double the individual charge of either C_1 or C_2. Regulation is dependent on the storage capabilities of the voltage-doubling capacitors and also on the rate at which their energy is withdrawn with respect to the replenishing rate of 60 Hz per capacitor.

For low-current loads that present a steady drain on the supply, lower value capacitors can be used than would be the case for heavier loads that fluctuate. For light-load circuits, capacitor values as low as 10 to 20 μF may prove satisfactory, whereas for more demanding loads, both the doubling capacitors and the filters may have to approach the 100- or 200-μF values. Since such power supplies are obviously economy types, it begins to be impractical to use excessive filtering, since the cost of the 500-μF or 1000-μF units becomes a significant factor, particularly at the higher voltage ratings needed to minimize burnouts. If greater regulation is needed, it would be preferable to use the full-wave types with a transformer delivering the required voltages rather than to use the doubling principle.

8-5 VOLTAGE TRIPLING

The line voltage (or the voltage from the secondary of a transformer) can be tripled by using three diodes plus three charging capacitors, as shown in Fig. 8-6. Here, when an alternation of the ac input voltage is positive, terminal T_1 is positive and T_2 negative. Electrons that leave the lower terminal T_2 seek a return path to the positive terminal T_1. Thus, diode D_1 has forward bias applied to it and will conduct under such polarity conditions. During conduction, capacitor C_1 charges to the peak input voltage with a polarity as shown. (The charge may be somewhat less than the peak voltage under load conditions, since current drawn from the supply diminishes the charge. For explanation purposes, however, we shall assume peak-voltage charges.)

During the next alternation, when T_1 is negative and T_2 positive, reverse polarity bias appears across D_1, and this diode will not conduct. Diode D_2, however, has forward bias applied to it and hence will conduct. Now, electrons leave terminal T_1, flow through D_2, and use capacitor C_1 for the return path to terminal T_2. Hence, the charge already present across C_1 adds to the input voltage and places a charge across C_2 that is proportional to the peak of the input voltage plus the potential already across C_1. This occurs because capacitor C_2 is across the two voltage sources and the potential appearing across C_2 is equal to the sum of the voltage across C_1 plus the ac line voltage.

During the next alternation, when terminal T_1 is positive and T_2 negative, the charge across C_1 is replenished, but rectifier D_3 also conducts at this time and charges capacitor C_3. When D_3 conducts, the charge across C_3 consists of the line voltage plus the existing charge across C_2. Because C_2 holds the *double voltage,* this adds to the line voltage, and the result is a triple voltage across C_3. For the circuit shown in Fig. 8-6, the lowest voltage can be obtained from across ca-

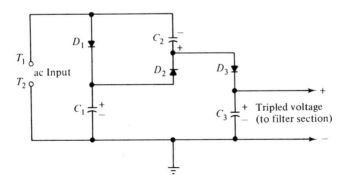

Figure 8-6 Voltage tripler

pacitor C_1. Twice the charge can be obtained from C_2, though in this instance the negative potential would not be at circuit ground. A voltage output of three times the line voltage is obtained from across C_3, as shown. The output, as with other rectifiers, contains pulsating dc, which must be filtered for a smooth dc by appropriate filter capacitors and series resistors, as explained in earlier sections. (If any of the other capacitors are tapped for a lower voltage output, filtering must again be utilized.) The filter factors given in Sec. 8-4 apply here also.

8-6 VOLTAGE REGULATION

Regulation is a term used to designate the changes in voltage output of a power supply under no-load and full-load conditions. The amount of voltage output obtained from a power supply depends on the voltage output of the secondary of the transformer winding, less the voltage drops that occur across rectifiers, series resistors, or filter inductors, if such are used. Because the amplitude of the voltage that develops across such components depends on the amount of current flowing through these units, the output voltage is also affected by the amount of current consumed.

With a minimum of current drawn from a power supply, the output voltage will be near the maximum value that it is capable of delivering, because the voltage drops across the series units will be at a minimum. As more current is drawn from the power supply, the voltage drops for the series units also increase, filter capacitors discharge, and thus the output voltage is reduced. Such a variation of output voltage with respect to the amount of current drawn from the power supply determines the voltage regulation.

Some load circuits tend to place a constant drain on a power supply; hence regulation is good, and no special means must be utilized for improving voltage regulation. In many cases, however, current demands on the power supply may vary, such as is the case with an audio amplifier, where changes of input signal amplitude cause the currents to rise in the power transistors. Thus, voltage regulation is poor, particularly if an input filter capacitor is employed. Poor regulation occurs because peak-voltage charges placed across the capacitor occur at short time intervals because the voltage peaks of the pulsating dc have a short duration. Consequently, as more current is drawn from the supply, the charge on the filter capacitors is drained off more rapidly than it can be replaced by the current peaks of the pulsating dc applied. With full-wave rectification, as shown in Fig. 8-3, the peaks occur more often and hence help to improve regulation. Also, larger value filter capacitors can be employed; in some special-purpose power supplies, filter inductors are used instead of a series resistor to aid in improving regulation. Such filter inductors are known as *choke coils*. With a choke input to a filter section, the voltage peaks are reduced to a lower level, and consequently there is less of a decline from the peak value to the average value. A typical full-wave power supply using a choke-input filter is shown in Fig. 8-7. The choke can have a value from 5 to 10 henrys, and the wire must have sufficient diameter to permit the proper amount of current to flow through it. Resistor R_1 is the series voltage-limiting resistor, though this can be replaced by an added choke for increased filtering in special applications. With chokes and transformers, special shielding precautions must be observed when such units are utilized near sensitive amplifiers or other circuits, because of the magnetic fields produced. Often the transformers or chokes must be angled differently to minimize stray field pickup.

The percentage of voltage regulation can be expressed by the following equation:

$$\% \text{ V regulation} = \frac{\text{no-load } E - \text{full-load } E}{\text{full-load } E} \times 100 \qquad (8\text{-}1)$$

Equation (8-1) relates the proportion of voltage increase or decrease with a change of load on a power supply. (The load on a power supply also includes any bleeder resistor network placed across it, though usually such a bleeder resistor consumes much less current and power than the actual load, such as a radio, TV receiver, amplifier, etc.)

As an example, if a solid-state amplifier's power supply has an output of 60 volts with no load but has a decline to 30 volts under full load, the regulation is

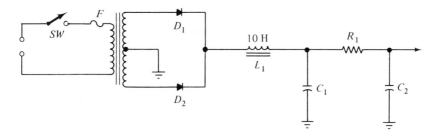

Figure 8-7 Choke-input filter

$$\frac{60-50}{50} \times 100 = \frac{10}{50} \times 100 = 0.2 \times 100 = 20 \text{ percent}$$

From Eq. (8-1), it is obvious that the greater the difference between the full-load and no-load voltages, the poorer the regulation of the power supply. Because good regulation in a power supply improves the performance and efficiency of both RF and audio amplifiers, some voltage regulation method is often used in well-designed electronic systems.

In some specialized industrial-electronic systems, where regulation is an important factor, choke-input filters sometimes consist of the *swinging choke* type, which is designed to produce an appreciable change in its inductance value with a change in the current flow through the choke inductor. The swinging choke is also known as a *saturable reactor* because it has a change of reactance for a change of its magnetic density when operated around the saturation levels of the hysteresis loop. The swinging choke has a high level of inductance when there is low current flowing through it. However, when current through the inductor rises, output voltage from the supply would tend to decrease, but the greater current flow through the swinging choke causes the magnetizing force to rise to the saturation point or near saturation, and the inductance value drops (sometimes from a 10- or 15-henry value to 5 henrys). Thus, the series impedance decreases and the voltage drop across the choke becomes less, to compensate for the voltage decline from the power supply. As the voltage drop across the choke declines, more output voltage is produced. Voltage regulation can also be improved to a considerable extent by the use of special voltage-regulating diodes, as explained in the next section.

8-7 ZENER DIODES

A solid-state voltage regulator unit that is extensively used is the silicon-junction *zener* diode. It has voltage-regulating ability because of the unusual characteristics that occur in the reverse-bias region. Here, as shown in Fig. 8-8, an unusual breakdown (zener) region prevails. At forward bias, it functions as a conventional diode, though with lower internal resistance and greater current-passing characteristics as the applied voltage is raised.

When the diode has reverse bias applied to it, its internal resistance is high initially, and only fractional amounts of current flow through it. As the reverse voltage is gradually increased, only a slight rise in internal conduction occurs. However, once the reverse bias reaches a

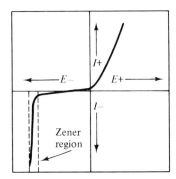

Figure 8-8 Zener characteristics

specific amplitude, internal resistance drops sharply and current shoots up, the value of which depends on the type of zener diode used and the circuit that it regulates. The sudden current increase during the reverse-bias breakdown voltage is shown in Fig. 8-8.

The feature of the zener is that regardless of sudden current changes within the unit, the voltage drop across it remains stable, making the device useful for voltage regulation. As the load draws less current from the supply, voltage tends to rise, and thus the current increases through the zener, bringing the voltage back to that specified. Similarly, if the load circuit draws more current, the supply voltage tends to drop, as does the current through the zener. Consequently, voltage regulation occurs again, and a constant voltage is maintained within the limit of the diode's zener region span. With zeners of more than one-watt ratings, heat dissipation may be a factor in operation, and heat sinks may have to be used, as was the case for silicon rectifiers or power-amplifier transistors discussed elsewhere in this text. Zeners develop most heat during the time when the load draws the least current. At this time the rise of power supply voltage increases zener current, and power consumption rises. With proper voltage adjustments of the supply, however, the zener will maintain the voltage at the value for which it is rated. For a specific circuit, it is important that the wattage rating as well as the voltage be considered.

The reverse-voltage breakdown does not damage the zener, as it would for other diodes. The breakdown occurs when a certain reverse-voltage amplitude is reached that is sufficient to penetrate the diode's internal semiconductor barrier, causing the unit to become a conductor in the reverse direction. After removal of the reverse voltage, however, the internal barrier region reforms itself without damage. The breakdown point can be closely estimated during design and manufacture by

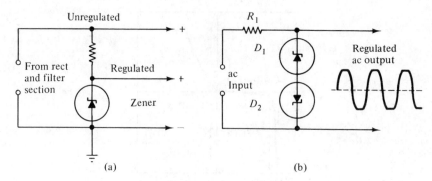

Figure 8-9 Zener circuits

control of the internal resistivity of the silicon structure. Thus, the zener point can be set at several volts or several hundred volts as required. In the ordinary silicon rectifiers not designed for regulation, the breakdown point is set at a sufficiently high value to be beyond the peak reverse voltage at which the unit is rated.

Two typical zener circuits are shown in Fig. 8-9. For the circuit in (a), note that the zener diode is placed in the circuit so that reverse bias is applied to it. Resistor R_1 must be selected so that the diode is held within the zener region for all fluctuations of load. Thus, the wattage of R_1 must also be sufficient to prevent overheating during the zener's current flow plus that of the load circuit. Thus, if the combined currents flowing through the zener and the load circuit equal 200 mA, this current flows through R_1. If the regulated output is 12 V, and the unregulated output is 15 V, a 3-V drop occurs across R_1 at a specific time. Thus the resistor would be approximately 15 ohms and though the wattage rating is indicated as approximately 0.6, a one-watt rating is preferable.

As shown in (b), zener diodes can also be used to regulate ac. Here two zener diodes are placed back to back, so that each half of the ac cycle is under control. Separate diodes can be used, or special double-ended zeners manufactured specifically for this purpose can be employed. With separate diodes, each should be tested to make sure that the operating characteristics are sufficiently close to minimize distortion of the output signal. As shown, the output waveform is clipped slightly at both the positive and negative peaks.

8-8 TRANSISTORIZED SHUNT REGULATOR

Transistors in conjunction with a rectifying diode can also be used for regulation purposes. A typical circuit is shown in Fig. 8-10. Here,

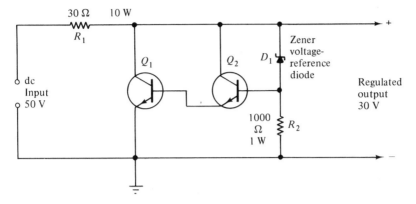

Figure 8-10 Transistorized regulator

transistor Q_1 is wired as a common-emitter amplifier, the base of which is directly coupled to the emitter of transistor Q_2. Essentially, these two transistors are in shunt with the output and hence the load system. The output voltage is regulated by virtue of the changing conduction of the transistors for any change in load current. This type of circuit also regulates the input voltage, and tends to stabilize it too by the regulation process. If the load circuit draws more current, the voltage tends to decrease. Hence, less current flows through the shunt network, and voltage across the series resistor R_1 decreases, thus raising the output voltage to compensate for the decline that would have occurred.

Such shunt regulators are sensitive to even slight E changes but are not very efficient. For the circuit shown in Fig. 8-10, for instance, an input voltage of approximately 50 would be required to produce a regulated output of approximately 30 V.

8-9 BLEEDERS AND VARIABLE-OUTPUT SYSTEMS

As mentioned earlier in this chapter, a resistor placed across the output of a power supply is often called a *bleeder*, because it "bleeds" some of the current and hence contributes to voltage stability. It acts as a small but constant load on the power supply and is one of the methods for improving regulation. Bleeders are designed to consume between five and ten percent of the current drawn by the load circuit.

As an illustration, consider (a) of Fig. 8-11. Here, the load for the power supply output draws 100 mA, and the bleeder has to consume 10 percent of such current (10 mA). Consequently, the value of the bleeder resistor should be 5000 ohms:

10% of 100 mA = 10 mA

$$R = \frac{E}{I} = \frac{50}{0.01} = 5 \text{ k}\Omega$$

The wattage in power consumed by the bleeder would be

$$P = EI = 50 \times 0.01 = 0.5 \text{ W}$$

Although the bleeder wattage rating is only 0.5 W, a one-watt resistor is preferred to keep temperatures below where they may damage the resistor.

When two or more series resistors are used across the output of the supply to produce intermediate voltages, the bleeder network is referred to as a *voltage divider*. Such a system is shown in (b) of Fig. 8-11 and consists of R_1 and R_2. This is the type of supply utilized in some industrial applications. As shown, two loads are in use, the first having 200 V applied to it at 10 mA, and the second load, 400 V at 190 mA. Design considerations here involve the current distribution between the two loads, and the common currents circulating through R_1.

As an example, assume that resistor R_2 in Fig. 8-11(b) has 10 mA of current flowing through it. As shown, however, resistor R_2 is shunted

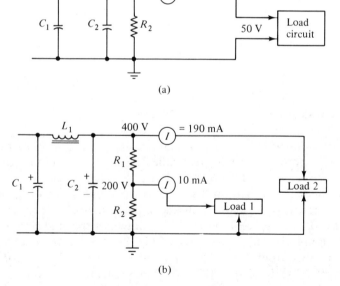

Figure 8-11 Bleeder factors

by the first load, which also draws 10 mA. Thus, resistor R_2 is in parallel with load 1, and the combination will draw a total of 20 mA through resistor R_1, since the electron flow for both R_2 and load 1 flows through this resistor.

Because there is a 200-V drop across R_1, and 20 mA of current through it, the resistance value of R_1 must be 10,000 ohms ($R = 200/0.02$). Resistor R_2 also has 200 V across it, and with 10 mA of current flowing through it, the resistance value must be 20,000 ohms. The first load circuit must also have a resistance of 20,000 ohms, since it also has 200 V across it and draws 10 mA. The resistance of R_2 and load 1 combined must, therefore, be 10,000 ohms. Obviously, both R_1 and the combination of R_2 and load 1 have equal resistance values in order to divide the 400 V exactly.

Several resistors could, of course, be used in series to obtain intermediate values of voltages as needed. Each load that shunts a section of a voltage divider draws additional current, and such additional current will flow through the resistors above it in similar fashion to the two loads illustrated in Fig. 8-11(b).

Instead of the fixed potentials available from a voltage divider resistor, there are occasions when a variable output is required. A simple series variable resistor can be used, as shown in Fig. 8-12(a). This replaces the usual fixed-value filter resistor between two filter capacitors at the output of a power supply. Here, resistor R_1 is selected to have the proper range for the variable voltage required, and also a sufficiently high wattage rating to minimize burnouts. If, for instance, the output from the supply is 20 V and a variation is required from 15 to 20 V, a 5-volt drop is needed when the full resistance of R_1 is in the circuit. The resistance required for R_1, however, depends on the current drawn by the load system. If the load circuit draws 50 mA, the resistance of R_1 must be 100 ohms, as shown below. Also, the wattage rating is indicated as 0.25 W, but a half-watt resistor is preferred to keep dissipated heat at a low level.

$R_1 = E/I = 5/0.05 = 100$ ohms

$P = EI = 5 \times 0.05 = 0.25$ W

There are limits to the variable resistor applications shown in (a), since greater current circulating through the resistor required a variable resistor of higher wattages. Wire-wound resistors tend to change value as the resistance unit heats up. A more practical variable supply is one utilizing a power transistor, as shown in (b).

In the circuit in (b), a single *pnp* transistor is utilized, with resistors

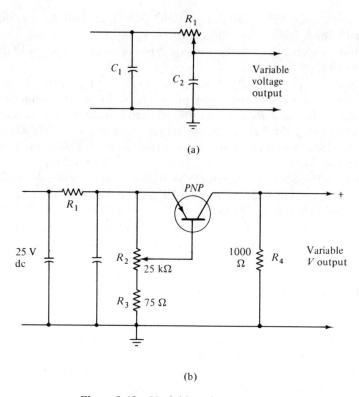

Figure 8-12 Variable-voltage systems

R_2 and R_3 shunting the plus and minus output terminals as shown. Resistor R_2 is a variable resistor of low wattage that adjusts the bias between emitter and base of the transistor, and thus regulates the amount of current flow between collector and emitter. Resistor R_4 is a bleeder resistor and also supplies collector bias in the absence of a load circuit.

The transistor must have an emitter-collector current rating above the amount to be drawn from the supply, and a heat sync is necessary. For a variable voltage control of approximately 25 volts, as shown, transistors such as the 2N173, 2N2147, or 2N2869 can be used. For different transistor types the value of the base-bias variable resistor may have to be changed from that shown for smooth voltage variation. A 50,000-ohm potentiometer can be utilized; if the range selection is less than that after the circuit has been tried, it can be replaced by a smaller variable value. It is difficult to estimate the exact values, since the load circuit is a factor in terms of the current it draws.

8-10 SCR UNITS

The silicon-controlled rectifier is a diode with a special gate terminal so that power can be switched on as required. Generally, the units resemble the type shown in Fig. 8-13(a), with a threaded terminal (anode) that is bolted to the chassis, with the flange acting as a heat sink as with the silicon diodes used as rectifiers, or the power transistors. The SCR units come in various sizes as needed for handling a specific amount of power. Units that measure approximately 0.5 inches wide and 1.5 inches high can handle currents up to 20 A at over 400 V. The symbol for the SCR is shown in (b), though the symbol shown in (c) has also been used on occasion. (See also Fig. 1-4.)

The silicon-controlled rectifier can be a conventional power-supply rectifier if simply the cathode and anode terminals are used. The presence of a gate electrode, however, puts the unit in a special functional category, where a small potential of negligible power applied to the gate can control many amperes of current flow and hence power applications to devices as needed. Consequently, SCR units find wide application in industrial control, automation processes, and in many

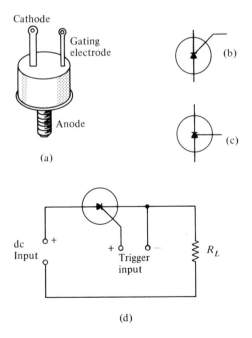

Figure 8-13 SCR unit and symbols

other areas where the amount of power to a load must be applied only at specific intervals.

When dc is applied to the input, as shown in Fig. 8-13(d), the rectifier is normally an open circuit and no current flows through the load R_L. To trip the gate function of the SCR, a trigger voltage is applied between the gate electrode and the cathode as shown. Such trigger voltage must have a polarity at the gate to coincide with the positive alternation of the input voltage. Once the SCR is gated, current continues to flow through the unit and the R_L even when the gating potential is removed.

Thus, the SCR latches on when triggered and remains in a conduction state until the anode-cathode voltage is removed (or the polarity is reversed). Since the gate potential is ineffectual once the SCR is triggered, the gating signal can be a short-duration pulse or a longer duration signal of proper polarity and sufficient potential to initiate the triggering.

When ac is to be controlled, the triggering mode is determined by the relative phase of the gate signal in relation to the anode signal. Because the gate control is lost once the SCR conducts, the use of an ac signal at the anode permits the *gate to regain control.* This is so because successive alternations of ac to the anode cause conduction to be interrupted at a rate coinciding with the frequency of the ac signal. This operational mode is shown in Fig. 8-14(a), where single-polarity pulses are used for triggering purposes.

Alternating-current signals can also be used at the gate. The operational mode for this is shown in (b). When the gate ac signal is in phase with the ac power input signal, the SCR conducts for successive positive alternations at the anode. When the gate signal E_g is positive at the same time as the anode potential E_a, rectifier current I_r starts to flow as soon as the anode and gate potentials reach values to cause conduction. When the gate signal is negative, the phase concides with E_a and conduction ceases. Thus, the SCR behaves as a half-wave rectifier.

If there is a 90-degree phase difference between gate and anode voltages, as shown in (c), the SCR does not start conducting until the gate voltage reaches a value to permit conduction, even though the anode has a positive polarity of voltage applied to it initially. When the anode potential drops to a specific point, conduction ceases, even though the gate signal may still be positive.

With a greater shift of gate potential, as shown in (d), the interval of conduction is still shorter, and in consequence less power is applied to the load circuit. Thus, by shifting the phase of the gate potential with respect to the signal at the anode, a considerable variation of output power is possible. The pulsating dc produced across the load circuit can

SCR Units / 241

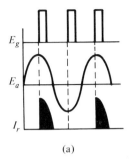

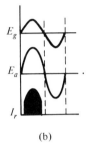

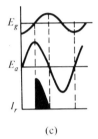

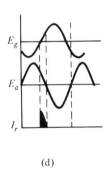

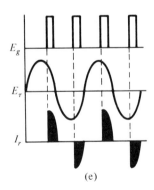

Figure 8-14 Thyrister operational modes

be filtered, if necessary, to reduce the ripple component and provide for a substantially pure dc, as was done with the power supplies described earlier.

The Triac is actually a pair of SCRs with inverted characteristics and hence is a bidirectional unit permitting current flow in both directions, as opposed to the unidirectional flow for the SCR. Both are, however, *thyristers,* having similar gating characteristics. (See also Sec. 1-4.)

As with the SCR, the Triac gate permits triggering and current conduction. Also, as with the SCR, once triggered, the unit is latched on and remains so until the current is decreased below the latch level. Representative pulse triggering is shown in Fig. 8-14(e).

Earlier in Fig. 8-13(b) a gate terminal was shown connected to the cathode section of a silicon-controlled rectifier. In Fig. 8-13(c) a terminal was shown connected to the anode section. When a single unit has both these gate terminals present (one from the cathode and one from the anode) it is termed a *silicon-controlled switch* with the symbol SCS instead of SCR (silicon-controlled rectifier).

The SCS can be triggered into conduction by either a positive gating pulse or a negative gating pulse, thus providing some degree of versatility. Also, the SCS can be triggered *off* by a gating signal, unlike the SCR which latches on when triggered into conduction.

While the symbols show the gate terminals connecting to cathode and anode, a four-layer *pnpn* junction semiconductor is actually involved, with the four regions tapped by the terminals. Thus, if a dc negative potential is applied to the anode, the SCS does not conduct. If, however, the anode has a positive potential on it, conduction is obtained by the application of a positive gating signal to the cathode gate, or by a negative signal to the anode gate terminal.

A typical example of SCR usage in television receivers is shown in Fig. 8-15, where it serves as a high-voltage limiter. For color receivers, high voltages in excess of 25 kV are used to achieve optimum brightness. If, however, the high voltage generated should become too high

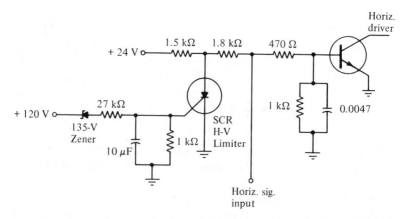

Figure 8-15 SCR as a high-voltage limiter in TV

(through maladjustments or defective components) some X-rays may be generated above the permitted levels established. Thus, most manufacturers provide some type of limiting device that prohibits function at excessive voltages.

The circuit in Fig. 8-15 is used in some Sylvania receivers (chassis E05, for instance). It comprises the horizontal sweep signal driver stage, using an *npn* transistor as shown. (See also Fig. 6-12.) The 120-V potential at the zener also relates to the high-voltage generating system, and if, by misadjustment of these potentials or because of defective components, this voltage rises, it approaches the 135-V breakdown potential of the zener as well as trigger voltage for the SCR. Hence, when an excessive voltage is felt at the zener cathode, the SCR is gated into conduction; the consequent low conduction impedance shunts the base input of the horizontal driver transistor and alters the bias. Thus, the driver transistor stops conduction, and hence the horizontal output system (and related high-voltage generation) will no longer function. Consequently, both the sweep system and the high voltage are turned off and remain off because of the latched SCR.

Shutting off the receiver will remove the gating signal from the SCR and will unlatch the horizontal and high-voltage systems again. If, however, the proper adjustments have not been made, or if the defects have not been remedied, the system will again cause a shutdown of high voltage.

APPENDICES

APPENDIX A

Resistor Color Coding

Molded-composition resistors have lead terminals at each end, as shown in Fig. A-1, and are available in various wattage ratings such as $\frac{1}{4}$ W, $\frac{1}{2}$ W, and 1 W, etc. The ohmic value of each resistor is indicated by standard color-coded bands grouped at one end of the resistor and is identified starting at that end, as shown in Fig. A-1. Usually four bands

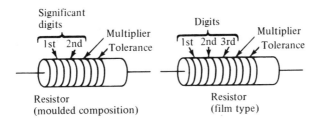

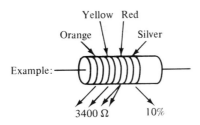

Figure A-1 Color codes for resistors

TABLE A-1: Resistance in Ohms (Ω)

Color	Digit	Multiplier	Carbon ± Tolerance	Film-type Tolerance
Black	0	1	20%	0
Brown	1	10	1%	1%
Red	2	100	2%	2%
Orange	3	1000	3%	
Yellow	4	10,000	GMV	
Green	5	100,000	5% (alt.)	0.5%
Blue	6	1,000,000	6%	0.25%
Violet	7	10,000,000	12.5%	0.1%
Gray	8	0.01 (alt.)	30%	0.05%
White	9	0.1 (alt.)	10%	
Silver		0.01 (pref.)	10% (pref.)	10%
Gold		0.1 (pref.)	5% (pref.)	5%
No color			20%	

of color are present for such resistors, and five bands are used for the film types. In either case, the last color denotes the *tolerance* that must be applied to the value obtained from reading the initial bands. Thus, if a rated 100-Ω resistor had a tolerance of 10 percent, its actual ohmic value could range from 90 to 110 Ω, for a variation of 20 Ω. A 1000-Ω resistor would range from 900 to 1100, for a variation of 200 Ω.

Table A-1 shows the color-code values and tolerances that apply to the carbon and film resistors shown in Fig. A-1. The abbreviation GMV indicates *guaranteed minimum value*. When a value is marked "alternate" (*alt.*), it indicates a coding that may have been used occasionally in the past but is generally referred to in modern components by the "preferred" (*pref.*) coding designation.

The standard listing for the resistor color coding is given in Table A-1, and a typical coding example is shown in Fig. A-1.

APPENDIX B

Capacitor Color Codings

The color codings for various ceramic-type capacitors are shown in Figs. A-2 and A-3. As shown in Fig. A-2, the tubular types may have *axial leads* (i.e., leads from the ends) or *radial leads* (i.e., leads connected at right angles to the capacitor length). The values read from the digit color are in picofarads (pF). A *temperature-coefficient band* is also present.

The coefficient of the ceramic capacitors is given in parts per million per degree Celsius (ppm/°C). A letter N preceding this value denotes *negative-temperature coefficient;* NP0 designates a *negative-positive-zero coefficient.* Thus, a designation such as N220 indicates a capacitance decrease with a rise in temperature of 220 parts per million per °C and denotes by how much the value changes during the warmup time. The NP0 types are stable units with negligible temperature effect on capacitance values.

As shown in Fig. A-2, five identification markings are used for these capacitors. For the axial-lead type, the identification begins at the end where the color bands are grouped. The first band represents the temperature coefficient marking, and the next two show the significant digits making up the value.

Often the axial-type ceramic capacitors have a wider first band for identification purposes. The radial types have an initial band of greater color area, as shown. For the five-dot disc type shown in Fig. A-2, the lower left color dot gives the temperature coefficient; the other dots in clockwise sequence use the same coding, as do the axial or radial types.

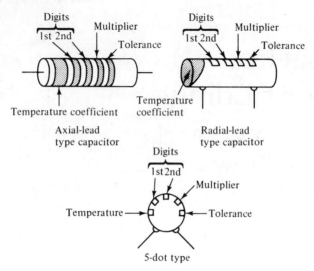

Figure A-2 Color coding for ceramic caps

TABLE A-2: Ceramic Types (Capacitance in Picofarads)

Color	Digit	Multiplier	10 pF or less	Over 10 pF	5-dot Temp. Coeff. TC	Extended Range	
						Significant Digits	Multiplier
Black	0	1	2.0 pF	20%	NP0	0.0	−1
Brown	1	10	0.1 pF	1%	N033		−10
Red	2	100		3%	N075	1.0	−100
Orange	3	1000		3%	N150	1.5	−1000
Yellow	4	10,000			N220	2.2	−10,000
Green	5		0.5 pF	5%	N330	3.3	+1
Blue	6				N470	4.7	+10
Violet	7				N750	7.5	+100
Gray	8	0.01 (alt.)	0.25 pF		*	†	+1000
White	9	0.1 (alt.)	1.0 pF	10%	**		+10,000
Silver		0.01 (pref.)					
Gold		0.1 (pref.)					

*General-purpose types with a TC ranging from P150 to N1500.
**Coupling, decoupling and general bypass with a TC ranging from P100 to N750.
†If the first band (TC) is black, the range is N1000 to N5000.

Capacitor Color Codings / 251

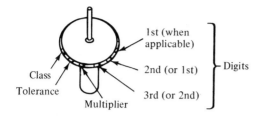

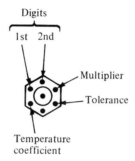

Figure A-3 Coding for button and feed-through ceramic caps

TABLE A-3: Mica Types (Capacitance in Picofarads)

Color	Digit	Multiplier	Tolerance	Type Classification
Black	0	1	20% (±)	A
Brown	1	10	1%	B
Red	2	100	2%	C
Orange	3	1000	3%	D
Yellow	4	10,000		E
Green	5		5%	
Blue	6			
Violet	7			
Gray	8			
White	9		10%	
Silver		0.01		
Gold		0.1		

252 / Capacitor Color Codings

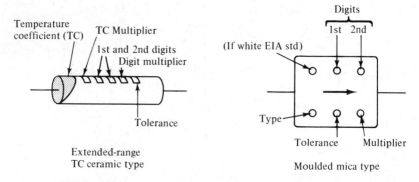

Figure A-4 Codings for extended range and mica caps

The button-silver mica, the button ceramic, and the feed-through ceramic types are shown in Fig. A-3.

The extended-range temperature coefficient ceramic capacitor and the molded mica type are shown in Fig. A-4. For the extended-range type, the first color sector indicates the temperature coefficient, as is the case with the five-dot types, but the second color segment represents the temperature-coefficient multiplier. Table A-2 applies to the five-dot and extended-range types, including the disc types. For those disc types that have only three color dots, the temperature coefficient and tolerance values are not given. In these, the first two dots (clockwise) are the significant digits, and the last dot is the multiplier, as shown for the button ceramic in Fig. A-3.

Table A-3 applies to the flat, rectangular-shaped molded-mica type of capacitor shown in Fig. A-4. An arrow, or an arrowhead, is imprinted on the capacitor face to indicate the direction of the color-coding sequence. The lower left-hand color dot shows the type or classification of the particular capacitor according to the manufacturer's specifications with regard to temperature coefficient, Q factor, and other related characteristics.

APPENDIX C

Transformer Color Codings

Color codings for transformers are shown in Fig. A-5 and apply to power transformers, RF and IF transformers, as well as audio types used in receivers. Although such color codes are generally accepted as standard markings, not all manufacturers follow them. Thus, for unknown units, testing is a must before usage.

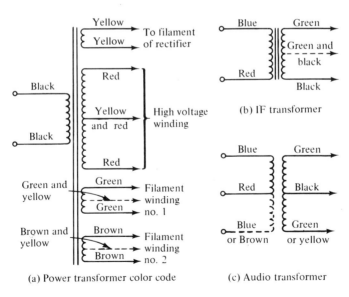

Figure A-5   Color codes for transformers

For the codings shown in Fig. A-5, black wires represent the primary of power-supply transformers, with secondary terminals marked as shown. The audio type shown in Fig. A-5(c) could be an interstage unit or an output transformer feeding the voice coil of a speaker.

APPENDIX D

International System of Units (SI)

The International System of Units (established in 1960) is based on the meter and thus represents a modernized version of the metric system. The official abbreviation is SI in all languages, though it is derived from the French phrase *système internationale*.

This system was adopted by international agreement, so it is the basis of all national measurements throughout the world, and integrates them for science, industry, and commerce. Thus, there is only one unit for a particular quantity, whether thermal, electrical, or mechanical.

The SI system can be considered an *absolute system*, using absolute units for simplification in engineering practices. Thus, the unit of force, for instance, is defined by acceleration of mass ($kg \cdot m/s^2$) and is unrelated to gravity.

The SI system is formed on a foundation of six base units of *length, mass, time, temperature, electric current,* and *luminous intensity*. Four of these are independent: *length, mass, time, and temperature*. The other two require the use of other units for definition, with multiples and submultiples expressed in the decimal system. Two supplementary units are also used: radian (for measurement of plane angles) and steradian (for measurement of solid angles). These two are termed *supplemental,* because they are not based on physical standards, but rather on mathematical concepts.

The standards for the six base units are those defined by international agreement. The prototype for mass is the only basic unit still defined by a rigid physical device. Thus, the kilogram is a cylinder of platinum-iridium alloy housed at the International Bureau of Weights

TABLE A-4

Quantity	Symbol	SI Unit	Derivation
BASE SI UNITS AND SYMBOLS			
Length	m	meter	
Mass	kg	kilogram	
Time	s	second	
Temperature	°K	degree Kelvin	
Electric current	A	ampere	
Luminous intensity	cd	candela	
SUPPLEMENTARY UNITS			
Plane angle	rad	radian	
Solid angle	sr	steradian	
DERIVED UNITS			
Area	m²	square meter	
Acceleration	m/s²	meter per second, squared	
Angular acceleration	rad/s²	radian per second squared	
Angular velocity	rad/s	radian per second	
Density	kg/m³	kilogram per cubic meter	
Electric capacitance	F	farad	$(A \cdot s/V)$
Electric charge	C	coulomb	$(A \cdot s)$
Electric field strength	V/m	volt per meter	
Electric resistance	Ω		(V/A)
Energy, work, quantity of heat	J	joule	$(N \cdot m)$
Flux of light	lm	lumen	$(cd \cdot sr)$
Force	N	newton	$(kg \cdot m/s^2)$
Frequency	Hz	hertz	(s^{-1})
Illumination	lx	lux	(lm/m^2)
Inductance	H	henry	$(V \cdot s/A)$

TABLE A-4 (*Continued*)

Quantity	Symbol	SI Unit	Derivation
Luminance	cd/m²	candela per sq meter	
Magnetic field strength	A/m	ampere per meter	
Magnetic flux	Wb	weber	(V·s)
Magnetic flux density	T	tesla	(Wb/m²)
Magnetomotive force	A	ampere	
Power	W	watt	(J/s)
Pressure	N/m²	newton per sq meter	
Velocity	m/s	meter per second	
Voltage, potential difference, electromotive force	V	volt	(W/A)
Volume	m³	cubic meter	

and Measures in France, with a duplicate at the National Bureau of Standards in the United States.

The meter is defined as a specific wavelength in vacuum of the orange-red line of the spectrum of krypton 86. Time has for its unit the second defined as the duration of specific periods of radiation corresponding to the transition between two levels of cesium 133. Kelvin for temperature is defined as 1/273.16 of the thermodynamic temperature of the triple point of water (the latter approximately 32.02°F). Ampere for current is defined as the current flowing in two infinitely long parallel wires in vacuum, separated by one meter, and producing a force of 2×10^{-7} newtons per meter of length between the two wires. The candela for luminous intensity is the intensity of 1/600,000 sq m of a perfect radiator at the temperature of freezing platinum, 2024 K.

Table A-4 lists the base units, the supplementary units, and the derived units. The derived units are produced without the use of conversion factors. Thus, a force of 1 N acting for a length of 1 m produces 1 J of energy. If this force is maintained for 1 s, the power is 1 W.

Index

Adjacent-channel traps, 116
Admittance, 34
Alpha (current gain), 27, 32
AM rejection, 136
Amplifier
    design equations, 70
    design variables, 68
    differential type, 177
    direct-coupled, 67
    IC type, 189
    load Z, 50, 77
    low-frequency type, 11
    modulated, 106
    operational, 179
    power load line, 51, 60
    push-pull, 62
    R-F type, 85
    stereo, 191
    tunnel diode, 8
    types, 11
    video, 118
Amplitude modulation (AM), 86, 106, 172
And circuit, 193, 196
Antiskate, 137
Automatic gain control (AGC), 15, 129
Automatic volume control (AVC), 172

Balanced modulator, 110
Bandpass filter, 113
Bandstop filter, 114
Bandwidth, 89, 92, 114, 119, 122, 135
Base, 11
Beta (current gain), 21, 35
Bias, 11, 17, 41
Binary, 193, 195, 217
    reflected, 218
Binary notation, 158
Bipolar transistor, 3, 11
Black box, 31
Bleeder resistor, 22
    systems, 235
Blocking oscillator, 149
Boolean algebra, 195
Bridge rectifier, 226

C-MOS units, 208, 215
Capture ratio, 134
Carrier signal, 85
Ceramic filters, 132
Channel, 14
Chip, 183
Choke coils, 231
    swinging, 232
Coefficient of coupling, 57, 121
Coincidence gate, 197

259

Collector efficiency, 55
Color codes
 capacitor, 249
 resistor, 247
 transformer, 253
Color killer, 170
Colpitts oscillator, 142
Common base, 23, 102
Common drain, 23
Common emitter, 11, 30, 35, 100
Common gate, 23, 38
Common-source FET, 18, 24, 37
Complementary circuit, 68, 208
Compliance, 137
Conduction band, 1, 3
Constant current, 182, 211
Constant-current analysis, 33
Constant voltage analysis, 33
Converter circuit, 142
Coupling
 coefficient, 57
 critical, 120
 factors, 119
 loose, 119
 tight (overcoupling), 120
Coupling ratios, 57
Critical coupling, 120
Crosshatch pattern, 156
Cyclic code, 218

Damping factor, 136
Darlington unit, 176
Decoupling, 24
 circuit, 64
Deemphasis, 112
Depletion MOSFET, 17
Design
 computer aided, 221
 prerequisites (switching), 220
 related equations, 70
 variables, 68
Detector
 AM, 172
 discriminator, 174
 FM, 173
 phase, 144, 170
 ratio, 173
Diac, 6
Differential amplifier, 177
Differentiator circuit, 165
 combination type, 167

Diffusion, 184
 isolation type, 185
Digital logic, 193
 polarity factors, 202
Diode
 junction, 4
 SCR, 239
 SCS, 242
 steering, 20
 types, 5, 9
 zener, 232
Diode-transistor logic (DTL), 201
Direct coupled transistor logic
  (DCTL), 206
Direct coupling, 67
Discharge circuit, 156
Discriminator detector, 174
Distortion
 FM, 134
 frequency, 47
 harmonic, 54, 134
 percentage, 181
Distributed capacitances, 46, 57
Doubling (voltage), 227
Down-state signal, 203

Efficiency
 collector, 55
 transformer, 56
Electromotive force, 3
Emitter, 11
Emitter-follower, 25, 35
Energy-level, 1
Enhancement MOSFET, 17
Exclusive-or, 216

Feedback, 20, 39
 current type, 67
 factor, 180
 inverse, 66
 loop, 179
Field-effect transistor (FET), 14
 circuit design, 76, 81, 104
 curves, 25
 design, 185
 IC formation, 185
 IF, 129
 parameters, 33
 transconductance, 26, 38, 79
 transfer characteristics, 27

Fifty-dB quieting, 135
Filter capacitors, 222
Filters, 113
    bandpass, 113
    bandstop, 114
    ceramic, 132
    crystal, 132
    IF, 132
Flip flop, 158
Flywheel effect, 98
Forbidden band, 1
Forward bias, 11, 17, 71
Four-terminal network, 30
Frequency distortion, 47
Frequency modulation (FM), 86, 108, 173
Frequency response, 136
Full-wave supply, 224

Gates (logic), 193
    combinations, 198
g_m (transconductance), 26, 38
Gray code, 217
Grounded collector, 25, 35
Guard bands, 86

h parameters, 31, 33
Half adder, 216
Half-power points, 91
Half-wave supply, 222
Harmonic distortion, 54, 134
Hartley oscillator, 141
Heat sinks, 5, 42
High fidelity, 133
High-voltage system, 174, 242
Hole flow, 3, 14
Horizontal sweep system, 156
Hybrid IC, 183

IGFET, 15
IHF sensitivity, 134
Image rejection, 136
Impedance (Z)
    factors, 31, 39, 40
    load Z, 50
    matching, 60, 77
    push-pull, 62
    reflected, 61
    resonance, 86, 99

Impurities, 3
Inhibiting circuit, 216
Input R, 36
Integrated circuits (ICs), 183
    commercial packages, 206, 215
    sockets, 189
    symbols, 189
    types, 188
Integrated injection logic (I^2L), 211
    nor circuit, 213
Integrator circuitry, 163
    combination type, 167
Intermediate frequency (IF), 129
    filters, 132
International system of units (SI), 255
Inverse feedback, 66
    current type, 67

JFET, 15
Junctions, 4, 11

Killer (color), 170

Large-scale integration (LSI), 184
Light-emitting diode (LED), 6
Light pen, 221
Load-line factors, 48, 50, 60
    push-pull, 62
Load Z or R, 50
    matching, 60
    power amplifier, 51
    push-pull, 62
Loose coupling, 119
Loss factors, 46, 56, 100

Memory, 162
Minimum error code, 218
Modulation, 86, 106, 108
Module, 183
MOSFET, 15
Multivibrator, 151
    one-shot, 161
Mutual inductance, 120

Nand circuit, 197
    MOSFET type, 205
Negative resistance, 7

Networks, 28
Neutralization, 101
Nor circuit, 196
Not circuit, 193

One-shot MV, 161
Operating point, 50
Operational amplifier, 179
Or circuit, 193
Oscillator, 106, 109
   3.58 MHz type, 143
   blocking, 149
   Colpitts, 142
   crystal, 138
   Hartley, 141
   multivibrator, 151, 161
   relaxation, 148
   tuner, 142
   variable, 141
   vertical sweep, 151
Overcoupling, 120

Parameters, 28
   h type, 31
   R type, 31
   Y type, 33, 37
Peaking coils, 118
Phase detector, 144, 170
Phase-locked loop, 144
Photodiode, 6
Pinch-off voltage, 79
Planar process, 184
Predistorter, 110
Preemphasis, 112
Power output, 54, 60, 135
   push-pull, 62
Power supplies, 222
   bridge type, 226
   full-wave, 224
   half-wave, 222
   high-voltage (TV), 174
   regulation, 230
   ripple reduction, 225
   variable voltage, 235
   voltage doubler, 227
   voltage tripling, 229
   waveforms, 225
Push pull, 62, 106

R parameters, 31
Radio-frequency amplifier, 85, 100
RAM, 162
Ratio detector, 173
Reactance, 86
   control, 145
Reflected binary, 218
Regulation (voltage), 230
   percentage, 231
   shunt, 234
Relaxation oscillator, 148
Resistor-transistor logic (RTL), 201
Resonance, 86
   ceramic, 132
   crystal, 132
   filters, 113
Reverse bias, 11, 17, 71
ROM, 162

Saturable reactor, 232
Sawtooth modification, 154
Schottky diode, 9
   clamp, 214
Selectivity (Q), 86, 93, 97, 114, 122,
   126, 135
Semi-conductors, 3
Sensitivity (IHF), 134
Sequential circuit synthesis, 221
Series feed, 105
Shunt feed, 105
Sidebands, 107
Signal-to-noise ratio (S/N), 134
Silicon-controlled rectifier (SCR), 6,
   239
Silicon-controlled switch (SCS), 242
Skin effect, 100
Small-scale integration (SSI), 188
Source follower, 25, 38
Special circuits, 158
Split stator, 107
   design, 123
Steering diodes, 20
Stereo amplifier, 191
Subcarrier burst, 144
Substrate, 9, 16
Swinging choke, 232
Symbols, 6
   diode, 6
   FET, 15
   letter, 69, 209
   transistor, 11

Terminal identification, 40
Thyristor, 6, 241
Tight coupling, 120
Time constant, 113
Tracking force, 136
Transconductance (g_m), 26, 38, 79
Transfer characteristics, 27
    forward, 34, 37
    reverse, 31, 34, 37
Transformer factors, 56, 60, 119
    push-pull, 62
    turns ratio, 61
Transistor
    bias, 11, 17, 71
    circuit design, 70
    curves, 22, 25
    Darlington, 176
    formation, 186
    heat sinks, 42
    terminals, 40
    testing, 44
    transfer characteristics, 27, 34
    types, 11, 13
Transistor-transistor logic (TTL), 200
    multiemitter, 203
Traps (TV), 116
Triac, 7

Trimmer capacitor, 143
Triode, 11
Tripling (voltage), 229
Truth tables, 195
Tunnel diode, 7
Turns ratio, 61

Unijunction (UJT), 13
Unilateralization, 102
Unipolar, 14
Up-state signal, 203

Valence electrons, 1
Varactor, 6, 131
Vertical sweep system, 154
Video amplifier, 118
Video signals, 110

Y parameter, 33, 37
Yoke, 156

Zener, 6, 19, 232
Zones, 4, 11